SCIENTIFIC INFORMATION AND SOCIETY

GYÖRGY RÓZSA

Scientific Information and Society

MOUTON · THE HAGUE · PARIS

Library of Congress Catalog Card Number: 79-184751

Printed in the Netherlands

Contents

List of abbreviations

CEUCORS	European Coordination Centre for Research and Documentation in Social Science (shortly: Vienna Centre), Vienna
CMEA	Council of Mutual Economic Aid
CNR	Consiglio Nazionale delle Ricerche, Rome
CNRS	Centre National de la Recherche Scientifique, Paris
CPSU	Communist Party of Soviet Union
DSE	Deutsche Stiftung für Entwicklungsländern
ECE	Economic Commission for Europe, Geneva
ECOSOC	Economic and Social Council of the United Nations
FAO	United Nations Food and Agricultural Organization, Rome
FID	Fédération Internationale de Documentation, The Hague
IAEA	International Atomic Energy Agency, Vienna
ICSU	International Council of Scientific Unions, Rome
ICSTI	International Centre for Scientific and Technical Information, Moscow
ICSSD	International Committee for Social Science Documentation, Paris
IFLA	International Federation of Library Associations, The Hague
ILO	International Labor Organization, Geneva
INIS	*International Nuclear Information System* (IAEA)
ISIS	*Integrated Set of Information Systems* (ILO)
ISSC	International Social Science Council, Paris
MARC	*Machine-Readable Cataloguing*

MEDLARS	*Medical Literature Analysis and Retrieval System*
MINESPOL	Ministerial Conference on Science Policy (UNESCO)
MSH	Maison de Science de l'Homme, Paris
NICER	National Institute for Social and Economic Research, London
NSF	National Science Foundation, Washington D.C.
OECD	Organization for Economic Co-operation and Development, Paris
ONG	Organisations Non-gouvernementales
R&D	*Research and Development*
UDC	*Universal Decimal Classification* (FID)
UNESCO	United Nations Educational, Scientific and Cultural Organization, Paris
UNIDO	United Nations Industrial Development Organization, Vienna
UNISIST	*World Science Information System* (UNESCO)
UNRISD	United Nations Research Institute for Social Development, Geneva
VINITI	All-Union Institute of Scientific and Technological Information (Russian), Moscow
WHO	World Health Organization, Geneva
WIPO	World Intellectual Property Organization, Geneva
WMO	World Meteorological Organization, Geneva
ZWO	Nederlandse Organisatie voor Zuiver Wetenschappelijk Onderzoek (Netherlands Organization for the Advancement of Pure Research), The Hague

Introduction

Even a few decades ago, a discussion on the social implications of scientific information would have seemed – to use a now fashionable term – 'futurology'.

Science was a matter for scientists, information on scientific literature for librarians, bibliographers and documentalists: the relationship between them was for centuries natural and almost undisturbed, government policy rarely interfered with either of them directly, much less with their mutual relations, or interrelations. Neither was there any mention of intergovernmental interest, of possible efforts at international settlement. Science and information on specialized literature existed on a national and international level within the framework of professional *corporative* bodies. Their social importance was formulated and played a very small part in government policies – that of the latter hardly ever.

With the passing time, with the emergence of the scientific and technical revolution, governments began to 'discover' science, formulating and expressing their socio-economic implications on a state policy level, in the form of *national science policy.* Since, however, science is by its nature of an international character (it has always been) and thus autarky alien to it, the emergence of international regional and then co-operation embodied in international organizations (UN and its specialized agencies) – starting from the national science policies – had only to wait a short time to become an *international science policy* on governmental level.

This development has been followed with *a time lag* by scientific information and the process is still developing. Its substance is that the problems of information which belonged earlier merely to the professional and 'corporative' sphere should become *structurally* an integrated element of national and later international science policies. The adoption by more than 80 countries of UNISIST, the 'World Science Information System' initiated by Unesco-ICSU, indicates among other things also the realisation by governments of information, in addition to that of science.

The time lag in development may be considered almost natural. As the scientific-technological and industrial development was followed by the recognition of the importance of *infrastructure* and its fitting into economic policy, in the same way information – which is one of the most essential elements of the infrastructure of science – became part of science policy with a time lag.

If the present trend of population increase does not change, from present forecasts, there will be about 6 thousand million people in the world in the year 2000.[1]

Natural resources and reserves such as food, energy, etc. will in all probability be unable to satisfy the multiple demands of rapidly growing populations, particularly in developing countries. Beyond *progressive social changes,* science is called upon to solve a major part of the task; this, however, necessitates the development not only of international science policy but of information policy as well. In this sense, not only science, but also information may be considered as a *natural resource of the nations.*

The above-mentioned necessity manifests itself also in other things as well. On the one hand, the investigations of the International Year Book announced by Unesco for 1972 report a high degree of *reading-hunger.* On the other hand, the Director-General of Unesco writes that growing specialization, the diversity of research results, their many facets, and the mass of knowledge produced by them threaten mankind equally with 'suffocation' and with 'paralysis' due to ignorance.[2]

The notion of a surfeit of information reminds us of the problem of 'environment' (biosphere, human environment). In both cases we have to deal with overgrowth. Human environment is endangered by the overproduction due to technical progress (air and other pollution, etc.), while the surfeit of information, the *information pollution* endangers scientific and technical progress itself.

As a remedy against both 'suffocation' and 'paralysis', research, and in its wake information, we employ computers. Their application is a new root for the intertwining of research and information. Computers, however, may be considered a 'remedy' or a stimulator of scientific progress only to the extent made possible by the *human element,* the intellectual potential. Namely, the information systems based on computers are not automatic, but *automated.* It follows that their positive endowments and their limitations are not merely of a technical nature but rather of social and human character. In this sense, too, scientific information may be considered a problem of social organization.

The socio-economic implications of science have been the subject of many books, studies and official documents. This is not quite so with scientific information. The time lag can be felt here too.

The title of this book,[3] *Scientific information and society,* indicates the effort at explaining the above-mentioned *tendencies.* It treats the interrelations of science and specialized literature, the governmental responsibility for both on national and international levels, the structural changes of information, and its growing social importance. It tries to outline – with the claim of a *synthesis* – the historical development of all these tendencies, their major socio-economic and science policy implications as well as their generalization by sociological and philosophical approach. This claim at synthesis is aimed above all at the formulation of problems.

CHAPTER I

The socio-historical evolution of scientific information problems

In his theories on the surplus value, Marx explains that the competence of the existing population is an invariable pre-condition of all production i.e. the principal accumulation in economy: this is the most important result of previous work but exists in live work itself.[1] This 'principal accumulation', that of production experiences, – knowledge systematized into science in the 18th century[2] – is being summed up and forwarded through and by the 'collective memory' of mankind – scientific literature.

This principal social function of scientific literature reflects the process of science becoming a productive force and is also one of the most essential parts of this process. Both sides of this process assume their final shape in the scientific and technical revolution unfolding in our age when the development of production must be preceded by that of technology and this in turn by that of the sciences.[3] Hence it follows that in the last analysis the material production of society is the technological application of the sciences.

A. THE PROCESS OF SCIENCE TURNING INTO DIRECT PRODUCTIVE FORCE AND THE PROBLEM OF SCIENTIFIC INFORMATION

The role of knowledge ('competence' of society) and its application in the development of the forces in production in these have evolved in three major historical stages,[4] within which takes place the transformation of scientific writings into specialized scientific literature, and with the increase of the latter, the appearance of scientific information and the concomitant problems.

In the first period, that is, up to the industrial revolution, we cannot talk about the social utilization of science as that is understood today. The organized, conscious and wide-scale application of science was not yet possible owing to the relatively undeveloped state of the productive forces. Science was studied individually, on a small scale, and the chief collective form of communicating was the book. Verbal communication and correspondence were also of significance, but that of the periodical publications was scarce. Scientists working without reference to one another were able to master the existing knowledge in their field, but this 'collective memory' was not disseminated beyond the walls of libraries.

The second period, lasting from the industrial revolution to the beginning of the scientific and technical revolution unfolding today, is, in fact, the first stage of an organized and conscious application on a social scale of science as a productive force. The sciences become specialized, new disciplines are born. Owing to the tremendous development of the forces of production, the isolated, individual scientific work is being replaced by scientific work organized on a large and wide-scale. As formulated by the Communist Manifesto: The intellectual creations of individual nations become common property. National one-sidedness and narrow-mindedness become more and more impossible, and from the numerous local literatures, there arises a world literature.[5] Beside the book, the journal becomes gradually the primary source of scientific communication. A direct survey of existing knowledge in a discipline by any single individual becomes impossible, a mediator, therefore, develops between scientific 'world literature' and research – first bibliography, then documentation.

The third period is the scientific and technical revolution itself in which science becomes a direct productive force. There is further specialization within the different scientific disciplines and at the same time there is a process of integration among them, as a result of developments in areas which overlap the lines of demarcation among the sciences. The natural sciences, mathematics, mathematical methods and cybernetics, in particular, are being used to an increasing extent in social sciences, especially in the concrete organizational-administrative branches (planning, management science, applied economics etc.). Work acquires, in general, a higher scientific standard, and science comes to embrace all walks of life. Scientific production and input increase *exponentially,* expenditures and the number of scientific publications doubles in every ten to fifteen years, research becomes the most rapidly developing 'industry'. Of all scientists recorded until the present time 90% have worked in the past decade and the number of research workers shows the most rapid increase in all professions. As a consequence of all this, the number of scientific documents is rising at a considerable pace; in addition to the books and periodicals, the research reports are gaining growing importance. Mediation between the literature and research (the library-bibliography-documentation complex) or, in a more general and collective term – *scientific information,* has assumed the rank of an independent discipline and professional activity in the social division of labour and has become, to a certain extent, an economic category. Viewed from the input side, the relevant expenditures in the world amount to some $ 500 million annually, according to international estimates.

The two essential aspects of the transformation of science into a productive force, are that the nature of the relationship between production and science changes, and that the social organization and application of science combines with production. Expressed in other terms: the development and application of science assumes organized forms and mass proportions and forms not only the basis of all production processes but itself acquires the character of a socialized production process. Science, in this way, becomes part of the production process not as a self-contained component, but as a relatively independent ingredient. It improves technology and production processes, it prepares the way for the introduction of radically new methods and for the creation of new industrial branches, and in turn, science is influenced by the production process. Production technique penetrates to the very heart of science, to the experimental work, imparting research itself a certain industrial character. This applies not only to the material equipment, to the manpower demand and to the large-scale organization in research but also to the direct products of research. For instance, isotope production in connection with research in nuclear physics, the manufacture of certain instruments or the breeding of hybrid maize, again shows the direct productive character of modern science.[6]

But within this integral unity, science enjoys a relative independence from production, which is of crucial importance in relation to its revolutionizing role. It is by virtue of this factor that science is able to draw attention, at a given moment, to seemingly abstract tasks which may not appear 'timely' or 'lucrative'. This relative independence applies not only to the choice of the subjects to be studied but also to specific methods of work, organization etc. The results of basic research in natural sciences, promoting both science and production, become applicable exactly by their general and abstract character (e.g. mathematics, cybernetics, mathematical logic etc.). These results gradually penetrate into social sciences, improving research possibilities and methods, promoting the development of new fields of research in the control, organization and registration of social-economic life (e.g. planometry). This process is subjected also to the dialectical interaction between natural and social sciences, an interaction asserting itself in two directions. On the one hand, Marxist philosophy – not as the science of sciences but as an instrument and method for the deeper understanding of the different scientific disciplines and of their relations to each other, – is gradually penetrating into the natural sciences. On the other hand, for the transformation of modern science into a direct productive force, the socialist mode of production may offer a more favourable soil. This, however, in no way implies an underrating of the scientific potential of the capitalist mode of production.

These sketchy disquisitions which do not seem to be directly related to our subject matter in fact set the background and give the key to an economic and scientific-organizational approach to the problems of scientific information, i.e. to their appearance which itself is a characteristic manifestation of the scientific-technological revolution.

Scientific information, as will be seen, shows that as a result of the scientific-technological revolution, certain activities that have evolved during the social division of labour will have to be given a new definition, as does the qualitatively new social function of science. Tentatively, the following can be said: *in the process of science becoming a direct productive force, the communication of scientific knowledge, itself a product of this process, reacts upon it and becomes an integral part of it turning into one of its most essential and active elements. Like science, it becomes organized, acquires mass proportions and tends to develop according to the organization of the productive forces.*

This development, following a tendency, has several components, its analysis has economic, science-organizational, gnoseological, and sociological aspects which require complex research. Just as by its very nature the process of the transformation of science into a direct productive force asserts itself most markedly in the natural sciences and engineering, so also, scientific information acts as an initiator of the productive forces mainly in these fields. At the same time, however, social sciences also undergo a significant development, especially with regard to the management of society and economy, in such branches as economics, sociology and management science. The application of mathematics, of mathematical logic, of mathematical statistics, of cybernetics, the introduction of concrete sampling methods and of models, attempts at, and experiments in, quantification are becoming more and more characteristic also of the social sciences. In this connection, scientific information in social sciences – to which it belongs – proceeds in the direction outlined above.

Thus scientific information has a two-fold significance: it is an integral part of certain areas of the social sciences which have a direct bearing on the development of the productive forces and it is a social-scientific activity affecting all the sciences. From this second aspect it follows, that scientific information has to be approached as a problem of the social sciences and according to economic and science policy categories.

Man himself is the basis of his material production, as of any other production that he carries on. All circumstances, therefore, which affect man, the *subject* of production, more or less modify all his functions and activities, and therefore also his functions and activities as the creator of material wealth, of commodities. In this respect it can in fact be shown that all human relations and functions, however and whatever form they may appear, influence material production and have a more or less decisive influence on it.[7] Thus scientific information could be shown to influence material production, to have a more or less determining effect through locating, classifying, forwarding and supplying valuable scientific information and through mediating between research and literature.

The 'world-literary' character of the literature, mentioned in the Communist Manifesto asserted itself at that time only as a tendency, but in the second half of the twentieth century it is a determinant of scientific development. This 'world-literary' character is an objective expression of the fact that autarky has no room in science, that 'national one-sidedness . . .' is becoming more and more impossible and that an international exchange of experiences has, instead, evolved in various forms as an objective necessity the most important of which being the specialized literature.

Parallel with this process has been the development by which work has come to require greater technical and scientific knowledge, which has set in motion a tendency towards *the intellectualization of labour.* Several factors are responsible for this development: the unprecedented rapid development of the productive forces, the social changes that have been unfolding all over the world and, related to these two factors, the increasing scientific and cultural demands and requirements of the masses. Within this general process, certain social activities, certain branches and processes of production have become considerably *research-intensive,* what is more, some sectors of material production are becoming directly dependent on research (e.g. atomic industry).

With the transformation of science into a direct productive force, with the coming to the fore of the actual organization and administration branches of the social sciences and of the more exact methods connected with them, with the internationalization of the literature and with the generally more scientific character of work, there has developed, as one of their sources and ingredients, the specialized literature. Specialized literature, and within that social science literature, has assumed tremendous proportions.

In a Marxist interpretation the task of this social-scientific literature cannot be other than '. . . to reveal the laws of the phenomena and thereby to provide information that can be used for the benefit of society, for finding one's way among the phenomena and for predicting their anticipated development.'[8] This requirement was energetically stressed and formulated in a concrete manner with respect to the various branches of social sciences and to the application of new methods of quantification at the General Assembly of the Academy of Sciences of the Soviet Union in 1962.[9]

Social science does not diminish but increases as a result of the technological and scientific revolution, and – let us add – so does that of social-scientific information parallel to it. This additional statement has a qualitative and a quantitative aspect, both being connected with the increase of specialized literature, with documentation in general, but it raises also specific social-scientific problems.

B. THE GROWTH RATE OF THE LITERATURE AND ITS CONSEQUENCES

Considering the three great periods of the application of science as a productive force, the growing volume of scientific periodicals tends to display the following trend (according to the data of the Auger Report published by UNO-Unesco):

Early 19th century	100
1850	1,000
1900	10,000
1960	100,000

If this rate of growth continues, the turn of our century will see some one million periodicals published.

As estimated by the Soviet Academician Dubinin, the number of scientific publications doubles every 10 to 15 years. According to the data of the American professor Kent, about two thousand books, newspapers, reports and other documents are turned out every minute of the twenty-four hours of a day all over the world, and the approximate volume of these publications runs into some 1,050 million pages.[10] This makes it easy to understand Professor Bernal's statement according to which *it is easier to make a new scientific discovery than to ascertain whether the given problem has already been solved.*[11] Again according to Professor Kent, research input in the USA in 1960 was 4.5 times that of the 1950 figure whereas productive research only doubled and, owing to the difficulties of access to the existing scientific data, 45 cents in the dollar are improductive, whereas according to Senator Humphrey's data, the ratio of 'overlapping' research attains 10 to 50% of all state-subsidized research activities in the USA.[12]

The situation in the social sciences is no more favourable. Although here the publications do not proliferate at the same pace as in the natural and technical sciences (let us remember the stream of technical-research reports turned out by the tens of thousands), in social science research – especially in social-economic organization and administration disciplines of a concrete, applied character, such as certain branches of economics, concrete research in sociology etc. – a significant part is played by the weekly and monthly publications whose documentation raises specific problems (e.g. demand for rapid, up-to-date processing, the relatively short-term utilization effect, the functions of the clipping archives).

Let us quote a few data on the increase in the volume of the literature relating to the social sciences. The Unesco Repertory of the social science periodicals of the world, omitting several branches (e.g. history, pedagogy) records 1,400 periodicals,[13] these being considered as the most important. The annual volume of the International Bibliography of Economics, contains, on an average, 7,000 book and article or periodical titles.

Thus the problem as to quantity is that information production is greater than the possibility of storing and retrieving while, according to international estimates, the amount of information doubles every ten years.[14] Information production has been treated in the literature already as an economic category, as in the review of the American Economic Society, where the author makes a detailed economic analysis of the *'information-producing firms'*, of the 'degrees of fabrication', making an attempt at methodologically formulating an input-output model of information production.[15]

As to quality, research in social sciences, like that in natural and technical sciences, must face complex and interdisciplinary problems. Almost any theme in social sciences embraces subjects belonging to more than one discipline. For instance, the problems of wage incentives are related to economic theory, economic policy (mechanism), planning, economic organization (wages), sociology, psychology, in a larger sense to pedagogy, ethics (morals), economic law (decrees) and last but not least to politics. *According to a survey, the majority of scientific studies covered one simple subject in 1900 but in 1950 some 85% of them deal with complex problems* (and this applies to the papers published in journals in particular).[16] The widening application of mathematical methods and of cybernetics – which itself is a product of complex research – also acts in the direction of increasing complexity.

An eloquent proof of this is deciphering the Maya manuscripts. The electronic computers in the Siberian Computing Centre of the Academy of Sciences of the Soviet Union, having been programmed for several months, turned out – within 48 hours – a complete dictionary of Maya manuscripts that had resisted all decoding until then.[17]

In connection with the above another problem arises. Owing to the development of scientific disciplines the lines of demarcation among them are becoming increasingly blurred. The research worker can be less and less certain that the articles of interest to him are going to appear in the periodicals in which they did in the past, when his speciality covered a narrower field. Thus, for instance, a study on the cyclical development of capitalist economy (a problem of political economy) that appeared in an American technical journal was then discussed in a Soviet journal on philosophy, which in turn was reviewed in a Hungarian political newspaper.[18]

While, on the one hand, periodicals specialized in compliance with the trends in research are being published in growing numbers in such fields as labour economics, market research, public relations or in such specialized border areas as economic law, mathematical economics, applied psychology etc., innumerable economic studies, on the other hand, appear in periodicals devoted to social sciences in general (this being the optimum case), to sociology, or in periodicals of a general content ('Trudy . . .', 'Sbornik . . .', 'Annales . . .', 'Proceedings . . .', 'Comptes rendus . . .', 'Sitzungsberichte . . .', 'Abhandlungen . . .', etc.), in annuals of scientific bodies, university or college annuals, in 'Festschriften', in mathematical, technological or agricultural periodicals. These periodicals of a general, non-specialized type reflect an earlier stage in the history of scientific literature but they have continued to exist and to appear in new forms.

These opposing trends in the world of scientific literature could be defined on the one hand as a *process of integration and of increasing complexity,* and on the other hand, as *a process of desintegration, of estrangement,* of fragmentation or differentiation. The specialization and integration process along which sciences develop, is reflected in the *content and structure of the periodicals* and in the literature in general.

In both their quantitative and qualitative aspects these problems appear also in scientific information, producing a mass of publications, which are difficult to survey, the more so, as these publications are also subject to the tendency of fragmentation according to areas of study and not only according to disciplines. As a result of which documentation ceases to be merely a problem of whether something falls into the domain of a main or of a subsidiary discipline. The problem arises that a given information may fall into the area of concern of several disciplines.

But alongside this tendency of fragmentation a trend towards integration also asserts itself. This raises the issue whether or not the present methods and forms of communicating knowledge are still apt to fulfil this function. Referring to John Bernal's suggestion made in 1948 at the Royal Society conference on scientific information, *to discontinue the publication of periodicals in their present form* and to have the scientists send their studies directly to one another, a French expert on documentation, Jean Meyriat, wrote that: 'the improvement of the present methods of information raises the question not only of the procedures of finding the data but also of those by which the documents are published.' It is becoming more and more evident that *the methods and forms of communicating information need a gradual and well-considered international regulation.*[19]

Among the quantitative, qualitative and organizational questions of communicating knowledge (forwarding of information on specialized literature) let us mention what might be called the 'points of gravitation' of the publications, which is also connected with the process of fragmentation. An example will be given with due regard to the pioneer work of P.N. Berkov.[20] The analysis of this example leads us to the registration-organizational problems of scientific information. A conference held at an Assembly of the Hungarian Academy of Sciences can serve as an example. Let us suppose, it deals with general sciences, and find out where such reports are published or registered or under how· many aspects one can come across them when seeking information.

The primary forms of publication or appearance are:

– as a report to the General Assembly, it will be included in the protocols of the Assembly;

– as the product of one of the sections of the Academy, it will be included in the report on the activities of the section;

– it may be published in the Proceedings of the section;

– it may be reprinted in separates.

The secondary forms of appearance are:

– in the bibliography of the academician as part of his publications;

– in the annual report of the author about his scientific activity;

– in one or more Hungarian bibliographies on social sciences;

– in international (Unesco and other) bibliographies on social sciences;

– in the analytical catalogues of libraries (special catalogues);

– the reprint may be included in the catalogues on bibliographies in libraries (author catalogues and subject catalogues);

– in the Hungarian National Bibliography, provided the material of the General Assembly appears later in a volume;

– in the Repertory of the Hungarian Periodicals;

– possibly in some foreign-language abstracting journal meant for information abroad.

But these are only the most obvious 'points of gravitation' for references. The secondary forms are not necessarily the result of extra work, but this does not detract from the tendency towards an increasing variety of references. In addition to the above thirteen variants, there are still other possibilities; when, for instance, the author publishes his report later in a volume of his collected works; this may then be abstracted abroad and included in various abstracts, etc.

Evidently, in a large number of cases, the time devoted to bibliographical and documentary registration, processing, selection, entry, classification, revision, bibliographical editing, abstracting, translation, indexing etc. of a publication *is comparable in order of magnitude* with the time devoted to its writing, i.e. with the intellectual input, provided the processing – as in our example – extends to various directions and on different planes. When surveying such an example one is tempted to quote the seemingly somewhat cynical statement of mathematician György Pólya, made to his students: 'The sources of a discovery are more important than the discovery itself'.[21]

C. THE RATES OF SPENDINGS ON SCIENTIFIC INFORMATION

The growth of the research apparatus on an international scale and the increasing input ratio assigned to research from the national income of various countries (2 to 3% in the advanced countries, totalling some $ 20,000 million, all countries included), the growing number of research workers and of their publications all over the world have considerably raised the sums allocated for scientific information.

The Council of Library Resources Inc. in the USA spent more

than $ 1.25 million on research during three years in the early sixties. According to international estimates, a sum of $ 200 million is spent annually on documentation and services in the world (for publishing directories and abstracts) and another $ 300 million are spent for storing, selecting and retrieving information material.[22] *The expenditure of $ 500 million a year on scientific information is in the same ratio to the annual $ 20 billion spent on research as the advanced countries expenditure on research is to their national income: 2.5%.* One single lubrication oil enterprise in the USA spends an annual sum of 50,000 on storing information data in IBM machines. According to the data of the National Science Foundation $ 124 million out of the $ 12 billion research and development expenditure went in 1963 to scientific information (more than to research in social sciences).[23] At the Washington Conference on Scientific Information, 1959, Otto Frank, breaking down the service spendings to its elements, presented the following data: simple bibliographical card (title and source), DM 1; same with annotation, DM 3; abstracting for machine storing and selection, DM 10 (the costs of copying and of the material used amounts to 10% of the input of intellectual work).[24] Hence, the prime costs of a classified bibliographical card processed with the traditional method and manually, with an annotation of a few lines, is about DM 4 to 5, which does not include the price and the handling costs of the publications (such as registration, storage etc.). The cost of a literature 'analysis' for a computer-assisted storage and retrieval programme is approximatively $ 2.

The annual operational costs of the Soviet VINITI (All-Union Institute of Scientific and Technological Information) were estimated at $ 5 million at the Washington Conference, but we also have reliable source data on its activities: 700,000 publications in the field of natural and technical sciences (partly also economic sciences) from 15,000 periodicals of more than 100 countries are processed annually.[25]

The picture becomes clearer on examining the data, advanced at the Washington Conference but not discussed at large, on the budget model of a 'world documentation' institute. Out of its assumed annual budget of $ 283,350,000 million, $ 49 million alone would go to translations. According to preliminary surveys, 3,400 abstracting services are operating at present in the world; if their yearly budget is assumed to be about $ 50,000 their sum total, i.e. $ 170 million is not far from the operational costs of an international documentation institute.[26] According to available data, the abstracting services of the world process only part – according to Bradford, only half – of the potentially utilizable literature, whereas the ratio of repetitions and overlapping abstracts is rather high – 2.7%.

It was to examine from the economic point of view, the effectiveness of information, that the UNDP (United Nations Development Programme) requested Jackson-report to do the 'Capacity study' concerning the UN technical assistance system.

The above few data demonstrate that scientific information, as was stated at the Washington Conference, can be looked upon as a natural resource of nations. A resource which has a direct impact on economic life, and the sums assigned to it, and in particular the sums assigned to documentation relating to production, can be considered as part of the production costs. Documentation has assumed an economic role, and, like research, presupposes an industry-like organization. As educational investments and input are, in the last analysis, production investments, since higher standards of specialization and general education directly raise productivity, the same applies to documentation, with the only difference that its direct or indirect effect on raising productivity and promoting economy depends on its application within the different branches of science.

Engels's statement on science expenditure can also be applied to the tendencies in scientific information. 'But in a reasonable state, going beyond divided interests . . . the intellectual element undoubtedly belongs to the elements of production and will find its place in the production costs. It is, at any rate, reassuring to know that the cultivation of science pays also materially.'[27]

D. TRENDS IN SCIENTIFIC INFORMATION RESEARCH

Scientific information, in order to fulfil its economic functions and to make 'its cultivation pay materially', has been submitted to a variety of examinations and requires even wider and deeper theoretical and practical studies. Yet the justification of such studies rests not on the large sums allocated to them, but on the achievements of scientific work that an efficient scientific information system facilitates. What is of importance, is not so much the expenditure of $ 500 million itself, but the many thousands of millions which depend on an effective application of information and therefore on an efficient system of scientific information.

Alongside the individual research projects the large scale organization research is gaining increasing importance. And this involves two noteworthy characteristics. First: a given theme requires *identical* material and intellectual inputs, including information from countries provided with different material, technical and intellectual resources. The difference will manifest itself in the *choice* of the research theme, because where the material and intellectual potentials are not available for the investigation of a given theme, research cannot be undertaken; the expense and capacities necessary for studying a given theme are identical everywhere. Second: the *developing countries,* their material possibilities and research potentials (including the intellectual ones) being more restricted, raise higher information requirements in many fields for they can only afford *reproductive* research. In this respect especially the countries that have embarked on the road of national independence display *extremely high information demands* extending beyond activities directed towards industrial development, to state administration . . . in social life. The modes of communicating scientific information do not develop in these countries gradually and organically from research: by virtue of the demand for them, they appear sooner than does research.

1. *Scientific information in the social division of labour*

Beside technical-development research, a variety of investigations covering several fields of scientific information are being carried on concerning such problems as the field of interest of a library (whether general or specialized), the co-ordination of acquisitions, the theory and methodology of bibliography, the structural aspects of the publications (indexing, author's abstracts etc.), the methods and forms of abstracting and so on. Two of these, bibliography and documentation, are in the forefront of contemporary research and as a result of their innumerable connections to scientific work certain problems arise which seem most suited to be tackled by social science investigations, and it is, of course, to social science that these two disciplines and fields of activity belong. When stressing these two fields of scientific information we are far from implying that a scientific library itself cannot be the subject of scientific investigation. In the social sciences there is a greater differentiation among their themes and therefore registration is at the same time a selection and a critical evaluation of the relevant works. It needs hardly any explanation to see that this *critical* approach in the social sciences is not identical with documentation in the natural sciences and engineering where the selection has a more objective character. It follows that the tasks of information in the social sciences are more complicated in this respect since aspects of ideology necessarily assert themselves in selection and communication; thus the thesis that there is 'no peaceful coexistence in ideology' manifests itself, *to some extent* also in information.

Hence the place of scientific information, and that of bibliography and documentation in particular, in the system of sciences should be investigated with an eye on this fact and with due regard to their role in the transformation of science into forces of production. More particularly, beside their function of general auxiliary sciences, it is important to find out in what scientific fields and to what extent they affect the productive force. Evidently, this can most effectively be analysed in disciplines having a direct impact on the development of the forces in production, that is – in the case of the social sciences – in economics, more exactly, in its applied, concrete branches.

When studying the *principal elements* of scientific information – a collective term including the most active fields of librarianship as well as the bibliographical and documentation activities – the following functions can be discerned: to locate, gather, select for use, and classify information irrespective of origin, time and language, to place the information data at the disposal of research by specific methods depending on how they are to be utilized, and finally to stimulate organically and regularly the relevant research demands. By doing so, scientific information saves research time, intellectual input and material means and affects scientific work directly in some fields and more indirectly in others. Scientific information in the course of the social division of labour evolved into a *self-sufficient* activity. It is an activity that saves live work directly and objectivized labour indirectly, by disclosing the literature (the documents) which *contributes to a rationalisation of intellectual labour and to increased material production.* By its inherent logic, the most active elements of scientific information – as a scientific activity – are the bibliography and documentation. Another important aspect of scientific information is that it is invariably attached to *some discipline* but evolves its specific requirements and metholodology taking into consideration its own general methodology. That is why the theoretical and concrete methodological research in scientific information is, at the same time, specialized scientific research, and also complex research since it is concerned with the general theory of scientific information which presupposes the involvement of several or all branches of science.

2. *Investigating the methods of communicating scientific knowledge*

Scientific information by having its own method of operation investigated *extends* its area of research from the secondary to the primary source. In addition to examining the modes in which its own methods, services and publications help scientific work, it contributes to the revelation of whether the primary forms of communicating scientific knowledge (books, periodicals, research reports), upon which information itself relies, meet research demands in every respect (the mode of publication, structure, editing techniques etc.).

The serious hint that the scientific periodicals in their present form have become obsolete and that they should be replaced by a direct distribution of scientific publications among scientists, has come from the side of scientific research. This problem ('l'avenir des périodiques scientifiques') has since Bernal's cry of alarm in 1948[28] been on the agenda in various forms without anything having been done in the matter: the number of periodicals has been increasing uninterruptedly and so has that of the abstracting services.

A new step in this field was the proposal of Academician Dubinin.[29]

Taking into account all the problems, besides the 'difficulties of growth' in the number of periodicals, Dubinin points out that the periodicals, even though their number doubles every 10-15 years, are unable to provide publication facilities, so great is the 'supply in articles'. In this connection the American Economic Review has published interesting data: in 1960 the number of the articles accepted and published attained 14 to 19% of those submitted for publication.[30] At the same time the boards of editors are trying to limit the volume of the papers whereby sometimes the descriptions of important original experiments and data fail to reach the arteries of the scientific world. Other factors limiting the usefulness of periodicals are the long time that elapses between the reception of the manuscript and its publication, the slowness of documentary publications (sometimes one year or even more elapses until a research worker learns of the existence of some publication of interest to him), the language difficulties etc. For all these reasons Dubinin has suggested, as a first step, to revise the coverage of periodicals – a task that he believes can more easily be achieved in the socialist countries than elsewhere, but of which he sees also the possibility of being realized on an international scale. The essence of his proposal is that it should be internationally enforced that publishers on accepting a manuscript for publication provide a summary, prepared by the author and consisting of three parts. The *paper itself* carrying the basic idea of the publication (restricted to maximum eight printed pages), *annotation* written in compliance with internationally adopted standards and with the specific requirements of the discipline in question, and the *research report* (the full material) containing descriptions, tables, statistics upon which the original work is founded. The annotation and the research report provided with the necessary code sign should then be sent to regional centres covering several countries where the annotations are published regularly in a conventional language and where – according to the demands received – the research report (the full article) is translated, reproduced and distributed. Thus the research workers will be informed of the works they are interested in at the time of, or prior to, the publication.

Scientific information can undertake different types of analysis in order to study and promote the further development of the methods of communicating knowledge. The investigations may extend to such problems as:

The relation of the subject treated in a periodical to its nominal area; the proportion of the major research themes within the journal; the degree in which the publications are covered by abstracting journals; how the literature related to areas bordering on several disciplines is divided among the different periodicals; the direction in which information production and literature increase within a branch of science, and within this the relative proportion of the different publication types (books, periodicals, research reports); the international function and role of the literature of a discipline in a given country etc. This method can be called the *reflective and analytical way of reviewing publications.*[31]

These investigations would provide only indications, and however exact methods are evolved, they would not permit us to draw ready conclusions. The point is that many subjective factors are involved which cannot be measured quantitatively, although their task is not this but rather to contribute to an objective evaluation of the problem mentioned above.

The same applies to the *structural-editorial analysis of the publications,* which is meant to help develop the most expedient principles in editing the 'apparatus' of the various types of publications, containing short introductory summaries by the author (at the head of chapters, for instance), various indexes (subject, author, geographical), abstracting etc. Although these elements do not add to the content, they greatly enhance the utilization of the publication. The importance of this question is also testified by the fact that it has been dealt with in various recommendations of different international organizations[32] and that definite attempts are known to have been made to regulate on an international scale the problems of social science documentation.

3. *Concrete research and survey-type investigations*

Investigations of this kind are being conducted and seem to be expedient to carry on in the following directions: the analysis of methodology used in contemporary social-science research, the analysis of the habits of the research workers in using bibliography, the analysis of the practices displayed in using documentation.

The documentary investigation of research methodology may eventually lead to the development of documentation methodology. The use of mathematical methods and cybernetics in certain branches, its effect upon research demands, the mechanization of documentation, the scientific information propedeutics, bibliographical introductions, the systematic arrangement of the research apparatus in the various disciplines, are the many problems to be studied in this connection.[33]

The relatively high number of such introductory manuals testify to the existence of *research demands,* on the one hand, and, on the other, adumbrate one of the main directions in social science research. The characteristic feature common to these works is that they do more than just register the bibliographical sources; they analyse their material from the point of view of application. Some go even beyond the classification of the themes (being a rather exacting form of documentation) and analyse the development and position of the disciplines, the research institutions and, in addition to the bibliographic and documentation sources, penetrate into the specific research methods of the individual disciplines. Acting in the reverse direction, these introductory manuals provide a survey of the literature and an introduction to bibliographical sources in order to help research methods. But whichever the starting point, depending on the aim, research in scientific information, underway or to be undertaken, should concentrate on *summarizing, describing, classifying and characterizing* such elements in each field of science as its specific research methods and institutions (scientific and practical organizations), basic periodical literature, information sources (bibliographical sources, library and documentation services, encyclopaedic and reference publications) and chronologies and statistical compilations, if such are necessary in the given field of science.

An analysis of *the customs of the users of specialized literature* by observation or interviews etc. could contribute to the clarification of the question as to what extent and in what manner the research workers use the literature in the different branches of science. Few significant surveys are known to have been undertaken in social sciences in this respect, which comes under research into demand – connected with the utilization of time, with the knowledge of languages and with other factors (which have been studied in natural and technical sciences) has been little developed methodologically in the social sciences. One of the more important surveys in relation to this question was conducted by Garvey, who was assigned to do this by the American Psychological Association.

Experiences accumulated and methods evolved in the natural and technical sciences[34] could be made use of to develop a research method in the social sciences. This is also a task which presupposes the creative co-operation of specialized research workers and of experts in scientific information. These investigations could cover such questions as how many hours per week, on an average, the research workers devote to studying the literature (according to types of publications), what is the reading-time ratio of published and unpublished documents (manuscripts), the extension of reading (whether they read only the publications immediately concerning their subject or peruse the entire periodical, including criticisms, information etc., to obtain a wider picture), the extent of literature used for research as compared with other information sources (talks, exchange of experience etc.), what would facilitate the use of the literature (author's abstract introducing the paper, indexes in books etc.). All this can be studied in connection with the *'research time budget'*. The investigations of this kind lead us up to another problem to be analysed and the two seem to be inseparable.

The study of the practices of the users of documentation should extend to questions such as: whether or not the research worker uses documentation and in what ratio to the primary publications; what types of services promote his work (forms of rapid information, such as observation files etc.); which form of abstracting seems best suited (such as short references or research surveys etc.); to what extent bibliography can be used in retrospective research; what is the estimated time saved when using documentation and bibliography. Such investigations are still awaiting answer in order to improve social science information. Some such initiatives, in the form of surveys, have been undertaken by the International Committee for Social Science Documentation.[35]

These questions of documentation 'marketing' cannot be evaded because if documentation is also an economic activity then it is necessary to assess the habits of consumption in relation to it. Scientific information *represents an economic value* (of differing amounts in different fields), and documentation is one of its chief transmitters, hence its transmission methods and forms should be measured for their efficiency. In certain fields scientific information represents merely an intellectual value but in these ranges too documentation saves time, and the demand for information must be investigated in these fields too, only the consequences will appear in a different form.

When investigating the practices in the use of documentation, one can discern *distinctive features* in the use of research methods and results and in the literature between natural sciences and engineering, on the one hand, and the social sciences, on the other.

The natural sciences and engineering have, for the most part, an experimental character within which basic, applied and development research can be distinguished. The same is the case in the direct applicability of their results which, in the last analysis, become essentially a productive force. The information data are of an objective character, the time factor (speed) plays an important role in the acquisition of scientific information. The results of the basic research are not affected by time, but on the other levels of research the durability of informations rapidly decreases. In development research the results can be directly measured economically. The processing of vast numbers of information data requires the use of documentation methods. The *'technical depreciation'* of a large part of information is extremely *rapid.* On account of the time factor, periodic publications and research reports are of decisive importance.

The social sciences have, for the most part, a verbal character, but the value of the results in the different branches again varies in practical applicability (historical sciences and humanities, on the one hand, and concrete organizational and administrative branches, on the other). The rate of the *'technical depreciation' is slower.* The information data have to some extent an ideological aspect. In general the demand for *retrospective* research is greater, and beside the documentation methods, the bibliography is also significant. The *information value of the books* decreases less rapidly, and among the periodical publications the weekly and daily press also represents a considerable reference source.

Although no strict demarcation line can be drawn between these two large fields of science in their use of documentation, the distinctive features enumerated above afford a certain basis for *differentiation.*

4. *Classification problems of scientific information*

a. *The systematization of sciences and the documentary classification*

The scientific and technical revolution of our time and its impact on society, the formation of a number of socialist countries following World War II, the mounting of the problems of the developing countries, the unprecedented extension of international relations and the new forms they assume, the economic and scientific-cultural competition between the two world systems, the extension of wide-scale application of mathematics to social sciences . . . all these factors have set more and more intricate tasks to social science research and to systematization and documentation.

The existence of the above-listed facts and phenomena and their reflection in the literature cannot be doubted. It is equally obvious that the concomitant concepts and notions assume different values and a new content; a host of new notions and technical terms crops up and their correlation and systematization raise numerous problems in documentation all over the world.

These problems acquire added intensity as regards documentation by the fact that the questions of classification, *systematization* and *terminology* in the field of the social sciences are of an ideological nature. However, this feature manifests itself in different degrees depending on the given field, branch and application. Obviously, the philosophical and gnoseological problems of science classification carry a greater and different ideological weight than, for instance, the structural problems of a scientific organization whose solution largely depends on considerations of practice, administration and organization.

At the same time there exists a close – though sometimes hidden – relationship between systematization, classification and terminology. Thus, for instance, the word 'science' in the English language means the 'exact sciences' and this reflects a certain theoretical position. Another example: socialist science organization generally includes philosophy or the science of history among the socio-historical sciences, whereas in western practice, e.g. according to Unesco terminology, philosophy comes under the category of 'human sciences'. The list of similar examples could be continued but they invariably show that science organization and science classification – especially in the field of the social sciences to which these questions themselves belong – will have to cope with many a difficulty.

The final solution of these philosophical and systematization problems is still far ahead, but in certain applied fields requirements of *practice* allow no delay in divising some kind of classification system. One of these applied fields, and an important one at that, is that of librarianship, documentation and bibliography, to be referred to as *documentary classification.*

The task of documentary classification *is not identical* with that of science classification, yet it cannot be independent of it either. Documentary classification is not called upon to systemize sciences but is expected to sort out *the documentary reflections of natural and social phenomena in the literature,* with due regard to the scientific foundations as well as to the specific methodological requirements of classification: making a differentiated distinction between the large numbers of publications, the classing of publications with a heterogeneous content and of those related to areas which are the concern of more than one scientific discipline, and therefore requiring to be recorded under more than one heading; providing the necessary cross-references as well as grouping them according to various auxiliary aspects, such as time, place, type of publication etc.

A publication, for instance, dealing with the 'peaceful economic competition' of the two world systems with special regard to a comparison of the productivity of labour in the industries of two countries – and there are many such studies – should be made accessible to research by classing it comprehensively and with regard to several aspects such as the socialist economic world system, the capitalist economic world system, the economic competition between them, the economy of the one country, the economy of the other country, labour productivity in one industry, labour productivity in the other industry – to stress only the most important elements of content in classification.

However perfect a classification of sciences may theoretically be, it can obviously not be called upon to meet all the requirements of a classification from so many aspects. Such a classification may be the task of a documentary systematization of an *applied character,* and a theoretical starting point to its creation could be furnished by a classification of sciences based on firm ideological and historical foundations. Both Marxist and bourgeois social sciences have made considerable efforts to develop the theory of science classification and to devise an up-to-date classification system for documents.[36]

The problems of documentation itself have extended beyond their own field and, in compliance with its economic and scientific significance, have come into the focus of interest of various disciplines. This fact has been given a concise formulation in the Wirtschaftswissenschaft periodical on economics of the GDR: 'Uniform classification is the basis of the much required economic documentation'.[37]

After research and experiments conducted for decades, the volumes of tables of a new classification system – not meant for international use – started to appear in the Soviet Union in 1960.[38]

Thus the social demand for an up-to-date documentary classification derives, in the first place, from practical necessity, a necessity of developing a 'common language' allowing access to the publications within a logical system elastic enough to assimilate new notions which arise uninterruptedly. In this formulation the documentary classification is not more and not less than *an instrument,* a scientific aid to research.

b. *The 'vocabulary' and 'grammar' of documentary classification*

The documentary classification as a scientific aid, whose primary field of application is the library and the documentation must have a 'vocabulary' or 'stock of notions' embracing all terms of social (and natural) phenomena, currently used and adopted in the literature, irrespective of their origin and interpretations. In the documentary classification of social sciences the notions of the universal 'vocabulary' can and do fall into two chief categories – Marxist or bourgeois ideology – *in the course of the application, interpretation, and concrete classification* of the individual cases.

We shall try to make this clearer by quoting two examples in the field of economics. 'The principle of distribution according to work in socialism', 'abstract work', 'concrete work', 'wage-fund planning' are such Marxist concepts of political economy which require no explanation in the socialist literature. However, in the western literature they are not used. In the socialist countries they are necessary for classing literature which relates to these concepts. And the other way round: the notion of capitalist economy is often expressed in the West by using such terms as 'free enterprise', 'freie Marktwirtschaft', 'welfare state' etc., and these cannot be dispensed with in the vocabulary if the relevant literature is to be assigned properly in western classification practice. It is again another question whether the literature dealing with these notions can be classed in the socialist practice with the notion of modern capitalism.

Hence, *the 'vocabulary'*, the 'stock of notions' is of *a general* character, *the interpretation* having *a particular* (specific) nature. The thesis of the old archivists, saying *Quod non est in actis, non est in mundo* should be modified in documentary classification to mean that only what does not occur in the literature can be absent from the 'stock of notions'.

Continuing on the linguistic level, the activization of the vocabulary requires a 'grammar', and the language being a conventional one, orthography is of particular importance. This is where we come across the greatest problem of documentary classification, especially in the field of the social science. The vocabulary can be regarded as both settled and in the making, life and practice are constant sources of enrichment, yet there is no universal theory, a 'descriptive grammar' underlying the *structure* of the 'language'. The systematization and classification of sciences would in fact be the grammatical foundation of this language; this would constitute the *basic research* on the basis of which the concrete *application* of the vocabulary could be extended to the field of documentary classification and scientific information. Yet in research it often happens that applied or development research is faced with the necessity of finding some solution before usually long-lasting basic research can arrive at some result.[39]

And this is what happens in the field of documentary classification; documentary classification practices are being carried on without a universal and detailed theoretical scientific classification, yet it should be emphasized that the former are not the simple functions of the latter because documentary classification *is governed by its specific inherent laws and methodological requirements which are in compliance with its purpose.*

c. *Documentary classification (in the social sciences) and the Universal Decimal Classification (UDC) – or the calculated compromise*

The use of the documentary classification has a really universal and international 'grammar' in the UDC for social science, which suggests certain instructive conclusions. We shall not dwell upon the debates of many decades about the UDC, its well-known advantages and drawbacks, but restrict ourselves to examining its usefulness for social science classification and, in this connection, the East-West co-operation in the FID C/3 (Fédération Internationale de Documentation) Commission concerning the revision of class 3, social sciences.[40]

It is nevertheless necessary to point out a well-known positive aspect of the UDC for social science classification. Owing to its comprehensiveness covering all sciences and to its notation techniques (the variety of relating aspects), the UDC can now express notions of considerable complexity which belong to areas bordering on several disciplines. This is a capacity of considerable importance as contemporary research tends not only toward specialization but also towards *interdisciplinary* investigations.

It is, now, not worth detailing, that the basic division of the system was almost anachronistic at the very time of its creation, that one of its almost 'incorrigible' drawbacks, especially in our days, is to set aside a whole entry for religion and to pack into one single class (6) all the applied sciences etc., and it can certainly be said without exaggeration that its *basic division does not correspond in content to any modern conception of scientific classification.* But in spite of this, owing to its advantages in practice: international character, universality, elasticity to extension, possibilities of combination, auxiliary tables etc., the use of a revised and reformed UDC seems to be the most suitable solution in social science classification, until further research can develop a really contemporary and scientifically founded documentary classification.

The experience obtained during the work of the FID C/3 Commission seems to show that such a revision can be actually achieved and that within *certain limits* and with certain compromises the utilizability of class 3 can be increased.

The UDC has its logical structure – reverting to our linguistic considerations, has a considerable 'vocabulary', suitable for assimilating new notions or extending the existing ones, has its own (obsolete and difficult) 'grammar' and has an (easily modifiable and modernizable) orthography. Coming on top of this is an important practical aspect: international experiences of some decades in the use of the UDC have accumulated, facilitating the assessment of the neuralgic points and helping to rectify the shortcomings.

The extension of the 'vocabulary' and the 'orthography' of the UDC thus permits modernization. An other modernization tendency of the UDC's is the research concerning its applicability for a mechanized retrieval system.[41]

The use of the UDC in the field of the social sciences – and also in general – can be considered as provisional until the conditions of switching over to another international system are established.

The use of the UDC as an aid, is a *calculated compromise* justified by its advantages in practice and by its limited but real possibilities of improvement.

5. *Research on technical development, mechanization and on thesauri*

The transition from traditional manual information methods to large-scale mechanized methods is *one of the most important trends* in research on scientific information. Experiments have been started first and foremost in natural and technical sciences since this is where they seem to be most justified economically. In addition it was here that manpower problems first appeared, on account of the volume of work, and made the most pressing demands (scientific information having the character of a product here) and it was this area, by virtue of the type of knowledge they involve, that was best suited for such experiments (the data being exact and quantifiable). Along with these two objective causes a subjective factor was also involved: the information experts in these fields felt technical development closest to them.

It follows that – under the pressure of economic and technical necessities – experiments to mechanize documentation were started in the natural sciences and technology, but its application in the social sciences is less developed, or more precisely, less widespread. These experiments – disregarding their detailed descriptions and their possible applications in technical development including reproduction techniques of extreme importance, microfilm-recording etc. – have so far been directed essentially to storing and retrieving data and to what is inseparable from these, to classifying and coding, which constitute the basis of all machine programming. The major trends in this new documentation technique are as follows:

– sorting with the punched-card system (marginal punched cards, usual optical systems) with manual, or electronic technique;

– systems based on microphoto technique, relying on coding, with photo-electrical selection;

– systems based on magnetic tapes or cylinders ('magnetic memories'), and its three modes of combination.[42]

According to Soviet data, the last-named method permits the 'storing' of 30 million pages of text within one cubic metre.[43]

American and Soviet experts agree that the constructional facilities are available to building computers of maximum capacity for storing and retrieving documentary data, but to find the most efficient method of programming and coding still requires much experimentation, particularly with relation to the social sciences. To be more specific it is not so much the maintenance of the machine itself which is expensive, but the continuous *intellectual preparatory work,* the elaboration of the *'thesauri'* and *analysis of documents* etc., necessary for a computer assisted storage and retrieval system. The analysis of one document (and this has then to be programmed) costs approximately 2 dollars. The average work rate of an analyst is 200 analyses per month (taking one month as consisting of 20 working days). If this unit is multiplied by hundreds of thousands, and it is on such scale that it is worth dealing with computers, the costs run into millions. The high level of automation however does not economise in manpower because such an information system needs an increasing volume of intellectual work.

It is precisely because of the high costs and of the relative shortage of specialists that a number of research projects in this field is carried out by international co-operation, under the aegis of international organizations: for example, the Euratom system, the INIS (International Nuclear Information System), or, in the realm of the social sciences (economics), OECD, FAO, ILO, UNIDO, Maison de Science de l'Homme, Deutsche Stiftung für Entwicklungsländern ('documentary pool') joint information system, and the ISIS (Integrated Set of Information Systems), see Appendix A of this volume.

The solution of the future, despite some still outstanding problems (lack of finance and specialized manpower etc.), promises to be in the direction of automatization, not because of the possible eventuality of reducing the cost and saving manpower, but because it could provide *a more efficient* information storage and retrieval. This therefore makes *no financial saving that can be directly measured but it can increase the productivity of research and development work which justifies the automation.* The research which is required today demands *a high degree of centralization* both of material and of intellectual capacities on the national and on the international level.

Thesauri: Beside the UDC several systems of classifications are also in use (e.g. Library of Congress System); certain information institutions have adopted their own classifications which are not internationally accepted. At the same time wide-scale research is being conducted to develop the thesauri for computer assisted information storage and retrieval system for social sciences. Thesaurus means 'a controlled list of terms, with indication of conceptually associated terms, for use in information retrieval systems, in connection with post-co-ordinate indexing'.[44] According to an expert of this question, '. . . the most concrete inference that can be drawn is that there is still immense work to be done: research is to be conducted, the necessary data collected and systemized, hypotheses advanced which then will have to be confirmed by computation and submitted to practice . . .[45]

E. ATTEMPTS AT INTERNATIONAL CO-OPERATION WITH SPECIAL REGARD TO THE UNISIST

The 'internationalization' of the specialized literature gives rise to problems of its international transmission, of its need for international cooperation.

Autarky should not be aimed at in the field of social science documentation.

The most important international co-ordinator of the scientific information exchange policies is Unesco. This role is fulfilled partly directly and partly through its assistance to non-governmental professional associations (ONG). In this field its international co-ordinating activities are numerous, ranging from regulating the international exchange of publications to such major projects as the UNISIST, a world-wide cooperation system of information, which is conceived jointly with ICSU (International Council of Scientific Unions). The two most significant international professional associations in this field assisted by Unesco are IFLA (International Federation of Library Associations) and FID (Fédération Internationale de Documentation).

Organized co-operation has been developed within the CMEA (Council of Mutual Economic Aid) in the field of documentation concerning technology, construction and agriculture. In the field of building industry the documentation was extended as far as a specialized processing of the foreign literature by the participating countries. The CMEA has founded the International Centre for Scientific and Technical Information. The Scientific Information Service of the OECD conducts significant activities in the information field and, for the purpose of aiding the developing countries, has established a specialized Development Enquiry Service.

International governmental and non-governmental organizations discussed the problems and possibilities of co-operation in documentation, i.e. this problem was discussed by and large in February 1963 by the Geneva UN World Conference devoted to questions of technical and scientific support to the developing countries. The secretary general of the ICSSD (International Committee for Social Science Documentation) made a report to the Conference on the problems of social science documentation in the developing countries.[46]

In February 1963 the experts of the UN Economic Commission for Europe discussed the problems of the abstracting services and the exchange of documentation in applied economics.[47] Referring to the resolution of the ECOSOC and to the recommendations of the UN General Assembly the report of the Executive Secretary of the ECE to the meeting says '. . . that such an exchange of documentation has a considerable bearing upon economic development and international economic co-operation'. The report reveals the overlappings and shortcomings in the documentation of the applied economics, and stresses the waste in the dispersal of efforts. It posed the question of what, in fact, belongs to applied economics (this question of classification arose, very characteristically, in connection with scientific information), and the tentative answer suggested was to include whatever comes under the concept of the economics of production (essentially everything except for economic theory, economic history and the history of theories). The report has also determined the range of the users: individual research workers, research institutes, production, governmental and international organizations making use of the abstracting services for the formulation and the implementation of their policies. According to the report, the developing countries, for lack of original documents, can only use part of the abstracts in economic work. The report treats several methodological and technical questions, such as the desirable size for abstracts (not to exceed 150 words), the need to exchange experience among editorial boards of documentary publications as well as between the Bulletin Signalétique of the Centre National de la Recherche Scientifique and the Referativnyi Zhurnal of the VINITI. It makes suggestions on the publications that may be of interest in the areas of interdisciplinary concern and in applied economics, on co-operation between the economic abstracting services and the UN Centre of Industrial Development (transformed into UNIDO, Vienna) to help the developing countries, etc., and gives a description of the abstracting journals – sixteen of them – appearing in English, French and Russian. Finally the report requests the ICSSD to continue its investigations (e.g. the sending of proofs for compiling abstracts) and, within this, to compile a repertory on the institutions of applied economic sciences and on the relevant periodicals.

The International Committee for Social Science Documentation (Secretariat in Paris) is a body created to conduct investigations and compile repertories as described above.[48] The Committee co-operates with the International Social Science Council as its consultative body in questions of scientific information. The ICSSD has a treble task: to act as the study and co-ordination centre of the consultative bodies of Unesco and large international social science societies; to be a research organization conducting experiments in new methods for the improvement of services; to act as a bibliographic and documentation centre to produce information publications or to take part in their production. Its activity covers four disciplines: sociology, politics, economics, cultural and social anthropology. In co-operation with the international associations of these four disciplines and with individual countries delivering materials it publishes annually the International Bibliography of the Social Sciences.[49] These four series constitute the most important undertaking in international social science bibliography.

Some important initiatives have been taken on the level of international co-operation which, though not directly involved in documentation, are concerned with collecting and cataloguing information in a manner which contributes to the development of international co-operation. To this category belong the Scandinavian Scandia-projects (collecting) and the American-initiated MARC I and II projects (cataloguing)[50] and the international research efforts concerned with the automation of documentation, see D.5 point of this chapter.

The UNISIST experiment. The above mentioned few efforts at international cooperation may all be considered only as a prelude to the UNISIST world programme worked out over several years by Unesco-ICSU, and adopted in October 1971 by more than 80 government delegations (See: UNISIST, *Synopsis of the feasibility on a World Science Information System,* Unesco, 1971, 92 p., and UNISIST, *Intergovernmental Conference for the Esta-*

blishment of a World Science Information System, 4-8 October 1971; Final Report, Unesco, 1971, 61 p.).

With UNISIST world-wide efforts at cooperation in the field of information, initiated by the Royal Society Scientific Information Conference of 1948 and the International Conference on Scientific Information of 1958 in Washington, *have entered a qualitatively new stage,* the responsability of scientists for information is now the responsability of *governments.* With this, the problems of information, which have involved government responsabilities in the more advanced countries on national level even up to now, have become government affairs *on international level, in world-wide dimensions.* Thus, the questions of scientific information exceed, in a sense, the framework of libraries and documentation organizations (though these continue to be bases for international cooperation) and even of scientific research organizations (though even the responsability of the latter increases through UNISIST). The secretarial duties of UNISIST have been prepared and are performed within Unesco by the Division of Scientific Information which belongs to the science sector, in co-operation with the Department of Documentation, Libraries and Archives.

UNISIST is *the philosophy, the movement and the organization* of world-wide cooperation in scientific information. It is *the hitherto greatest experiment* in international cooperation, its scope extending to both primary and secondary publications, the 'information analyst centres' (a new type of information service) and the data-banks. The scope of cooperation includes the developed and the developing countries, the United Nations-family, intergovernmental and non-intergovernmental professional organizations.

UNISIST, as a philosophy and as a world movement adopted by governments is a major result, and as an organization it is a huge experiment, the evolution of which needs time and a whole series of measures of an organizational and financial character.

In the historical period when science turns into a productive force, the significance of scientific information increases, its economic role – within the social sciences, chiefly in the economic sciences – acquires added emphasis and develops into an *economic category.* Scientific information becomes a component part, an active element of this process. Thus its place in the social division of labour requires a qualitatively new formulation and a theoretical foundation, because scientific information itself – as part and parcel of science becoming entirely and directly a productive force – develops in the direction of becoming the category of the productive forces *through transmissions and adventitiously.* The analysis and the theoretical generalization of the specific features of this development can be achieved through a variety of research projects.

The basic problems of these investigations have grown now beyond the needs of individual research workers – be these for information, bibliography or literature. They are no longer the worries of single institutions or disciplines or those of scientific information institutions: they have become, all over the world, collective social-economic problems of science policy.

CHAPTER II

Scientific library in the age of the scientific and technical revolution

The main argument of this chapter may be summed up in the following headings:

– Owing to its scientific and economic importance, the scientific library by itself forms a particular field of research.

– The scientific information is an integral part of the information spheres in the intellectual communication system of society.

– The tertiary information function and the flow of information of special literature can be carried out by a hypothetic multi-channel model.

– The comparative backwardness of theoretical work is one of the main reaons for the virtual impossibility of resolving the contradictions between the individual tendencies in library policy and the different 'oriented' approaches, as

library-science oriented approach,
documentation-oriented approach,
history-oriented approach,
information science-oriented approach.

– Scientific library has two features: it appears as part of the general library system and as part of the entire body of science, and as such, is a subject of science policy.

– The theoretical formulation and foundation of the division of labour between universal scientific libraries (as general) and special libraries (as particular) is one of the key issues of a further development of scientific librarianship as a whole.

– Scientific libraries have particular tasks in reconciling the contradictions, divergent views, and problems arising from the 'two cultures' (i.e. the [natural] scientific vs. humanistic erudition and culture).

– Scientific libraries also have particular tasks in eliminating backwardness, in reaching the world standard in the economic, scientific, and technological fields, and in increasing the 'per capita well-being' in the humane field.

To be examined are the following questions. Whether the *practice* of scientific libraries (including special libraries and documentation) can rely on well-founded and developed theoretical works or studies as regards the conception of development? The requirements of the scientific and technical revolution, its place in the social division of labour and the interrelationship between its components? And whether there are some questions raised throughout the world by the scientific and technological development and economic growth, which relate to the general educational policy, sociology and library theory, in short: to *library philosophy?*

The answer to the first two questions is in the negative, while to the third in the affirmative. However, these questions are closely related, and the answers to be given and a thorough explanation of the problems involved is conditional upon a long-term, multi-dimensional, and institutional research work on the part of the specialists.

This study seeks to contribute to this research work by raising certain questions and outlining certain *hypotheses.* To raise the key issues of a debate may, in itself, be profitable in certain cases, and the question of the place and development of scientific library in the circumstances of the scientific and technical revolution is a key issue.

A. THE SCIENTIFIC LIBRARY AS A PARTICULAR FIELD OF RESEARCH

Theory is nourished by practice, and generalizes the practical experiences. At the same time – as exemplified by science – a good theory is ultimately the best practice. Starting from certain practical examples, which will serve to illustrate certain points raised in this study. To consider some data on the work of scientific libraries: the real social needs they are expected to meet; to what extent they have succeeded in meeting these needs; on what basis they are developing? Only a full knowledge of these problems may justify the posing of hypothesis. As a *case study,* let us take the example of Hungary, a country with a comparatively well-developed cultural and library system, and with minor and limited economic resources and possibilities.[1]

At the end of 1970, the number of scientific and special libraries in Hungary was about 800 (the statistical survey showed a figure of 1935, but this exaggerated figure includes also bodies where some information functions are carried out but which still cannot be considered to be libraries); of this, the number of libraries securing information in specialized literature at a national level is 94, those employing a full-time librarian is 511 (thus altogether 605) and the number of specialized services which can still be considered as libraries, is about 200. The 1935 libraries shown in statistics have some 28,6 million library units in their possession, the amount spent on new purchases was 186,5 million forint. The staff of the 605 scientific and special libraries was about 3000 (counting only the full-time employees; the figure is estimated). In 1968 the number of special translations recorded was 63 836, including 805 078 pages. In the same year, the number of loans was above 4 million, and library research ('Literatur-recherche') was conducted in 7181 subjects, totalling 163 447 items.

What *research apparatus* had to be satisfied by this *scientific library base?*

In Hungary there were 1046 bodies on December 31, 1969 where research and development activities were performed, of these were 131 research institutes. 2,51 per cent of the national income was spent on research and development, amounting to 6,4 thousand million forints. The number of those employed in R&D was 48 000.

In order to illustrate the size of the problem, we present side by side, without performing any statistical refinement, the data on the research apparatus and the scientific library basis:

	Research	*Scientific and spec. libraries*
Work places	1046 of this: 131 re- search institutes	800 of this: 187 in research and designing institutes
Staff	48 800	3000 (in the 605 libraries)
Inputs	6,4 billion	186,5 million
R & D subjects	23 300[2]	
Library stock and services		28,6 million units bibliographical research in 7181 subjects (involving 163 447 items) 63 836 special translations (805 078 pages)

The picture becomes clearer, if one takes into account that those active in R&D were 1,21 per cent of the total economically active population, and that of these 2923 people had scientific degrees.

Some of the data of the above table indicate conspicuous deficiencies. The 23 thousand subjects would be an improbable bad proportion if compared with the research staff of 48 thousand. As opposed to the 6,4 billion R&D inputs, the purchase of specialized literature amounting to 186 million seems to be low, and the relative proportion between libraries and librarians also seems unfavourable.

The main deficiency, however, is that the statistics of scientific and special libraries is not a part of statistics on R&D.

To be sure, these data need further specification and correction, but this is not necessary here; nor is it necessary to indicate certain ratio regularities or to make international comparisons since they intend only to support factually the statement that scientific library itself, by the significance of its tasks, is a subject involving complex investigations in library theory, science policy, technological development, and economics. These investigations should be highly differentiated according to library functions (promoting education, research, and technological development) and to the library's 'profile' (specialized fields of interest, branches of science).

Besides these practical requirements, scientific libraries as cultural institutions represent certain *'intangible assets'* which cannot or can hardly be expressed in quantitative terms. And it would have little meaning anyhow, since libraries form an organic part of a country's intellectual assets, like other public collections, major theatres, opera houses, etc. also indicate *the country's intellectual level irrespective of their practical usefulness at a given time.* At any rate, it should not be left out of consideration that the concrete and readily applicable results of scientific library work, even in the field of natural sciences and technology, may only *indirectly and through transmissions* make their effect felt.

Owing to its manifold functions and the social demands it has to meet, a scientific library, acting as an aid to scientific and economic work, as a workshop of higher education and training, as an institution for the diffusion of culture and general knowledge, and last but not least as a scientific institute, poses a complexity of scientific problems. This being so, i.e. that a scientific library itself is a scientific problem, it is evident that the solutions may be attained by an approach on a theoretical level and by scientific methods.

B. INFORMATION SPHERES IN THE INTELLECTUAL COMMUNICATION SYSTEM OF SOCIETY
THE DUAL ASPECTS OF DOCUMENTS

One important junction of questions of a theoretical nature concerning scientific libraries is the relation of the latter to special literature information. The term 'special literature information' will henceforward be understood as information given on and from special literature.

The socially necessary information on special literature is an *integral part* of the *socio-economic information system* including scientific progress and technological development. The scientific approach to their problems may be conceived in relationship with the *whole:* the information system, and the *part:* the information on special literature.

Society needs every kind of information, irrespective of *its provenance and form,* which may be effectively utilized within organized social activities, economic, scientific, technological, etc. What is strongly underlined here is the *content, the applicability, and the value of information* and not its channels and its techniques. The latter are also significant since they render information realizable. However, their treatment has been in any case widely discussed in the literature of librarianship.

The information on special literature is *a subordinate concept of the intellectual communication system of society.* The *specific significance* of information on special literature within the intellectual communication system of society largely depends on time, subject field, and on the purpose of application. Information on special literature in every respect forms only a part of information.

Information on special literature as has been referred to above represents a 'collective memory', the continuity of knowledge and intellectual assets, which involves information on special literature, both primary and secondary, i.e. its collecting, storing, processing, and transmitting institutions, the types of services which functionally and organizationally materialize in the historical sequence of their evolvement in library, bibliography, and documentation. Special literature information in this sense is *not identical* with documentation, in other words, documentation is *not synonymous* with the totality of organized information:

documentation → special literature information → information system.

Starting from the most general sphere this process may also be represented this way:

information system → special literature information { library / bibliography / documentation

Documentation forms the most mobile part of special literature information. It is concerned with that part of the entire body of knowledge which *is subject to 'technical depreciation',* i.e. to redundancy and is characterized by a *vast amount of data.*

Considering that the bulk of data is in part *readily applicable* in economic and technological development, more or less *ephemeral* in character, the *speed* of their processing and transfer is the most important factor. Attempts at solving the mechanization of documentation follow from these three factors.

The dual aspect of documents

Besides its content, form, and circumstances of its diffusion, special literature is also to some extent determined, by *the handling* of the actual documents. A price list in a company (in its calculation section) is an aid of daily use, while in a national library it is looked upon as a national publication, a specimen of 'museum value', or in a special library it is handled for research purposes. This poses the question of whether a statistical survey (published in 50 to 100 copies or 'non-published') designed for a restricted circulation is 'special literature'. And again, whether standards, patent specifications, prospectuses, catalogues of industrial fairs, market reports in foreign trade in a few dozens of copies for official use only may be regarded as 'special literature'.

In this sense, we may speak of the *double feature of documents*. In a broad sense, the categories mentioned above may all be qualified as documents. However, as regards its *content*, special literature can also be taken in a *narrow sense* which – feasibly – does not permit the arbitrary inclusion of the former categories in it. From the aspect of documentation, these 'non-traditional' documents (research reports, prospectuses, standards, etc.) belong to the sphere of special literature information or rather to the still wider sphere of economic and technical information.

The double feature applies not only to special literature. All information, necessary to society, also has at least two features. First it appears as information explored and transmitted to be applied to a certain task: '*direct* mission-oriented information', then it may also appear as information serving current awareness and general orientation: '*indirect* mission-oriented information'.

Information which lends itself to be used in production in economic or technological activities seems feasible to be treated as *a special product* and as something representing *economic value*, emerging from and serving the purposes of the process of reproduction on an increasing scale. In other words, a criterion for this examination should be formulated in terms of economic categories according to this process:

production → distribution (circulation) → consumption.

Here the problem is approached of special literature information from the 'consumption' or utilization side in contrast with the more customary approach which takes the 'production' side, implying a quantitative consideration of the subject.

What the 'production' – quantitative consideration of special literature information, a practice which has been predominant up to now – implies is that the stress is placed on the production of *secondary information*, that is, the products of documentation play the leading role. From this it emerges that the system of special literature information may well be compared with a *railroad without time-table* along the lines of which the trains are loaded at stations (products of documentation) and the traffic manager's only concern is to let the trains start out from his station. After the train's departure he does not mind any longer what is going to happen to it: whether it will 'collide' with other trains; what other 'parallel' freights are underway; whether or not it will safely reach its destination; where it will be unloaded – thus once the train left, he does not care.

It is therefore essential to shift the stress into 'consumption', on to the use of information, and on the basis of this to define its production and distribution, too:

the socially necessary special literature information → production of secondary information (with a reasonable division of labour) → distribution (circulation, according to needs).

C. A HYPOTHETIC MULTI-CHANNEL MODEL OF THE FLOW OF INFORMATION

'INFORMATION OFFICERS', OR THE 'TERTIARY' INFORMATION FUNCTION

Documentation does not exist by itself. It is always the documentation of *something*, of a branch of science, of a profession, of an art, etc. Documentation has to start out from the actual needs of these things – *even if the individual sectors are not always able to formulate their needs.* These needs may be *latent* and it is precisely one of the most important tasks of special literature information to contribute its own inquiries, theoretical and practical, to raising and formulating the needs.

All this may seem obvious or even commonplace, but these problems will appear in another light by taking stock of the individual products of special literature information, a good part of which will appear as being produced without adequate planning of 'market research'.

The *'hypothetic multi-channel' model* of the flow of secondary information – and of the organized special literature information in general – may be outlined by taking as an example a company or an institution which subscribes to several primary publications (reviews) and secondary ones (abstracting journals). One of these reviews may be called X_1 and one abstracting journal Y_1. The field of interest of X_1 and those of Y_1 are common (e.g. industrial economics).

X_1 carrier of primary information

Y_1 international abstracting journal (Western), of secondary information, which overlaps X_1.

Y_2 international abstracting journal (e.g. Referativny Zhurnal), another carrier of secondary information, which also overlaps X_1.

Y_3 abstracting journal of a national documentation centre, a third carrier of secondary information, which naturally also overlaps X_1.

Y_4 abstracting journal of the national documentation centre or institute of the given industrial branch, which, again, may not neglect the content of X_1.

Y_5 information bulletin, published by the company itself, which – besides Y_1-Y_4 – reviews, in detail X_1 as an important journal of the given industrial branch since the company knows best – and this is probably so – what it needs, and, after all, X_1 can be most rapidly and directly processed 'at home' and addressed to the 'desks', to the individual concerned.

> And now the circle closes: there comes the chief engineer or the chief technologist and requests X_1 the carrier of primary information, because no documentary publication can substitute the direct scanning of the important X_1.

It is highly probable that this hypothetical 'multi-channel' model is, by and large, characteristic of the flow of special literature information. This being so, the question arises whether it is not more feasible to adopt the much more efficient and economical method of buying only the best abstracting journal available and so avoid the duplication in the information processing. A similar process is already taking place in R&D, where patents, licences, 'know-how' are obtained on a contractual level to avoid unnecessary duplication. The scientific information can be considered as a part of the 'know-how'. In practice this would mean utilizing the most outstanding international abstracting journal rather than having an ineffective maze of secondary information in a 'multi-channel' system at a national level. The model outlined above would be 'demoralizing'. However the author's main argument is that the cause of scientific information should be treated on the basis of serious programmes elaborated with scientific methods, based without exaggerations on considerations of the real needs of science and the national economy.

'The information officer', or the 'tertiary' information function

It is necessary that the frequently mentioned 'information explosion' should be reformulated, be made more precise. The exponential increase of the specialized literature has brought in its wake the multiplication of secondary information. From this has followed that now it is almost impossible to find one's way even in the secondary information. *Too much information in the end makes available too little information;* the quantity does not become quality. In this respect mechanization (automation) can help only partially. Mechanization makes sense only if the input is of a very large quantity, but if it does make a selection of tens and hundreds of thousand items then the selected items will be too large to enable a perusal of the important literatura of one topic. This process brings to mind the problem of traffic congestion; as vehicles go at greater speed and supply greater comfort, they multiply to an extent where they obstruct each other and slow down transport.

There is a *congestion* also *in the flow of secondary information.* The problem is not that there is not enough information but that there is too much, which is not selected, sifted, and above all not identified for the purposes of individual users.

It appears therefore that whether the traditional or the mechanized information processing is used, it is necessary to introduce between the researcher and the information processor a new communication link, which could be referred to, as it is by now the practice in specialized literature, as an 'information officer'.

There already exist such research teams in which the exclusive task of one or two of its members is to provide the team with the required information on the basis of the primary and secondary sources of literature. They do this by finding the relevant information from the secondary sources and without making abstracts, classification etc. These 'information officers' are of the *same status* as the other members of the research team.

The role of the 'information officer' within an institute or research unit would be to analyse regularly and on par with the level of the research references, bibliographies, abstracts, and in general documentary publications, and notably to bring to the attention of individual researchers material which is relevant to their work. Another effect of the processing of secondary information would be to draw attention to new topics and in this way *stimulate research.* The present *maximum* of the processing work of documentation would be the *minimum* of the 'information officer's' processing work: apart from producing subject-syntheses it would not be concerned with the processing of the literature. To put it another way: the work of the 'information officer' *begins* where that of the documentalist *ends.* What is being considered here is a *tertiary information function.* This type of activity would be really an organic part of the research process, its most efficacious element. It would, apart from a knowledge of the library and documentation, demand a high level of specialized knowledge of the research area in which he operates.

The systematization of the functions of the 'information officer' that is suggested here, is at the same time a conscious formulation of *the tendency* that the users of documentation are *in the first place information specialists* and not researchers.

The schema of the flow of information

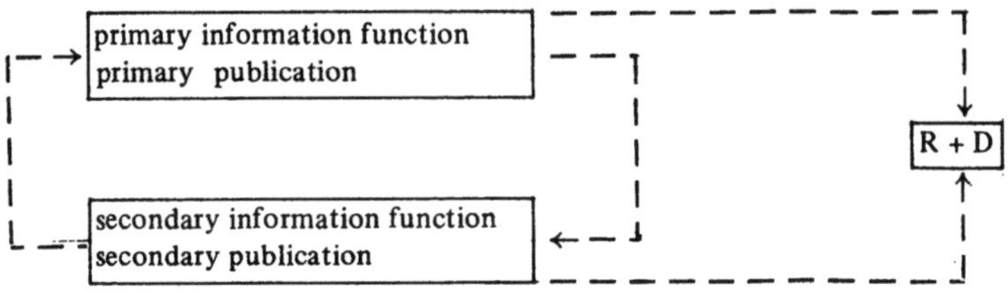

The schema of the flow of information with 'information officer'

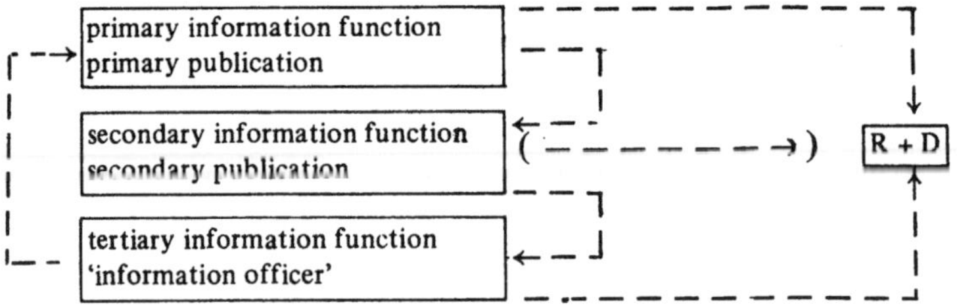

users of
- primary publ.: R + D + documentation
- secondary publ.: documentation + 'information officer'
- 'information officer' : R + D

D. THE INDIVIDUAL TRENDS IN LIBRARY POLICY AND THE DIFFERENTLY ORIENTED APPROACHES

In respect of theoretical work and studies the actual situation is far from being satisfactory though the number of publications on libraries and librarianship adds up to a separate library. For this there are many reasons, practical and subjective, one of them being that particularly since World War II librarians have had to face a vast amount of urgent daily routine work (storing, cataloguing, reference, and the related organizational, budget, and similar problems) on account of the information deluge and the increase in circulation. So much so that their energies and interests have necessarily been shifted towards the solving of pressing functional, operational, and organizational problems. Another factor affecting the comparative backwardness of theoretical work is what Marx said of science but which *mutatis mutandis* applies even more to the library problems, namely that the value of science as the product of intellectual work has always been underrated since the working time needed for its reproduction is not proportional to the working time needed for its original production. Thus, e.g., a schoolboy may learn the binomial theorem in an hour.[3]

Thus, librarians with scientific ambitions abandoned the not too promising field of library theory and tended towards one of the 'established' branches of science and scholarship where they were not exposed to indifference or to the danger of being qualified as a pseudo-science. It is quite another question that library theory may be successfully developed only in *close connection* with the specialized branches of science and scholarship the support of which forms the primary task and justification of scientific libraries.

Another hindering factor has been (particularly in the past) the often unfruitful debates, going on for reasons of prestige, between librarians and documentalists which, in fact, covered organizational and administrative problems for the most part. As a reaction to the sudden advance of scientific and technical documentation, and also because the scientific library could not readily respond to problems arising from the rapid development of science and technology (nor could it be prepared for it), libraries of humanistic studies and social sciences found their traditional attitude justified. This stressed, one-sidedly if not exclusively, the study of historical problems (history of books and libraries, processing manuscripts and old books, retrospective bibliography, etc.). This raised the question of whether scientific libraries need 'library scientists' or 'learned, scientific librarians', a controversy which is greatly responsible for the fact that library science, whose concept and scope have not been formulated clearly enough up to now, tends to give priority to well-established and widely accepted investigations into cultural history over library theory. It is obvious that what scientific libraries need are scientific librarians who have a creative proficiency in one or another branch of science or scholarship and are conducting research in it but who also pursue their profession with a good knowledge of the particular scientific methods and skills of library work.[4] What the concept of the 'learned or scientific librarian' implies is the acceptance of the individual librarian as a scientist or scholar but it also implies the underestimation of library work as a profession of a scientific nature, while the concept of 'library scientist' includes both of the former but – on account of its unclarified content – in a way that raises doubts about both of them.

All this and other reasons not mentioned here have led to a standstill in library theory, sometimes even to its devaluation, and to the development of virtually irreconcilable trends and differently 'oriented' approaches.

These can be outlined, by and large, as follows: *library science* (in close association with the public libraries) – *oriented, documentation-oriented, history-oriented* and *information science-oriented approaches.* Let us now examine their development, content, the problems of their irreconcilability in an effort to find possibilities for a synthesis.

1. *Library science-oriented approach*

This approach is founded on the traditions of historically developed large libraries from the times when scientific information as a concept did not exist and when specialization in science, and consequently in the library field, was still at an initial stage, when the general scientific library was the only depository of it and directly furnished literature to science. The intensive study and handling of highly valuable collections, the formulation of the place of library – to use a modern term – in the division of labour, and to some extent the theoretical generalization of the scientific library's practice are all factors which clearly explain the library science-oriented approach which has gradually developed into 'Bibliothekswissenschaft', 'library science' or 'bibliotekovedenie'. The appearance and the extremely rapid development of documentation and public libraries, owing to the unprecedented rate of scientific and technological progress and to the revolutionary changes in the public demands for culture and education, particularly in the socialist countries, have confronted libraries with a new situation throughout the world. *The relative homogeneity of librarianship* has disappeared, wide-ranging networks of special libraries and documentation centres have evolved, networks of public libraries have accounted for an ever growing share in the cultural budget, and society has shown an increasing interest in them. As a result of these processes, the scientific library has – virtually or sometimes actually – been pushed into the background, and documentation with its natural freshness and expansive methods has assumed a good part of the information functions or has developed new ones which the libraries had not been prepared to fulfil. Thus the scientific library seemed to have somewhat passed over to the defensive, and documentation with its new functions and methods appeared *not as a partner* created by scientific and technical needs but *as a rival.* The expansive work of documentation and its sometimes unaccountable attitude towards the scientific library (exaggerated emphasis on the library's conservativism and on independence of library work) have deepened the conflicts. In practice, all this appeared as quarrels over matters of competence, organization, and administration, and the relevant theoretical considerations had led to misunderstandings or often to debates which, however, today are declining in intensity. The other process, the immense development of public library system, in turn, has given a new impetus to 'library-orientedness' in another respect. The growth in the number of libraries and the development of internal library work gave rise to a variety of practical problems such as training, organization, co-ordination, which made it *imperative to collect experiences and organize their exchange.* This resulted in *methodological work* which, in turn, gave rise to theoretical generalizations underlying what is known as 'library science'. Thus 'library science', interpreted in different ways and understood to have different contents in time and space, concentrated essentially on *the internal* work of libraries, considering it as a scientific task – which it really is in certain respects – and in this endeavour the old library-oriented approach encountered an interpretation of 'library science', evolved from the new practice of public libraries.

The main reason for misunderstandings lies not so much in the label – after all why should there not be a science for librarianship if there is theatre or motion picture research? – as in the fact that debates of several decades have not succeeded in filling the concept with unambiguous content. Furthermore, several practical problems have also come under the concept of library science the solution of which, although requiring scientific training, cannot be looked upon as science.

By analogy: the surgeon operating on an appendix does not think himself, and is not thought of as, a scientist engaged in scientific work. The same applies to the engineer designing a building, and so on, and so forth. But no one would doubt that all these activities rely heavily on science, and these professionals must have scientific training and qualification. Likewise, it is also obvious that not only the study of the theoretical foundations of these activities is science but also the theoretical generalization of experiences gathered from practice. Or, taking a library example, classification as a library operation is an activity requiring scientific training but not science. The study of the theory of classification, in turn, may well be considered a science. The only question here is whether the theoretician of library classification is a scholar of 'library science' or his activities – a form of scientific classification – belong to the sphere of philosophy. However, what really counts is not the label but the content.

What has been said of the library-oriented approach as the most general collective term necessarily applies to certain elements of the other three 'oriented' approaches, a briefer description of them will therefore be sufficient.

2. *Documentation-oriented approach*

The objective reasons for the evolvement of the documentation-oriented approach are deeply rooted in the economic and technological progress. It was about the turn of the century that the needs of technology and practical economic work raised demands for new forms, methods, and content of information. This new-type demand was called forth by industrial companies, and – to a lesser extent – by science and it would be a mistake to believe that it was of a merely technical character. To be sure, the forerunner of contemporary abstracting journals, *Chemisches Zentralblatt,* started as early as 1830, and documentation archives were set up within the technological and development offices of industrial plants, it is also true that archives of economic documentation started their work in those early years within as well as outside libraries.[5]

This new-type information demand was characterized, first of all, by speed, by the many-sided presentation and analysis of the content of periodical literature in compliance with the customer's needs, and also by the processing of documents other than traditional publications (business reports, prospectuses, price-lists, and so forth). All this naturally involved the development of working methods and forms not peculiar to scientific libraries. Thus, for instance, one essential function of scientific libraries, the preservation of the holdings, is partly or fully absent from documentation. The literature, presented and analysed by documentation, is not necessarily available at documentation centres whose main task is not to preserve the source material but to supply information from it.. This, in fact, is its chief peculiarity. However, it should be added that no theoretical consideration is against uniting or combining these two basic types of information services within one and the same institution as proven – and also disproven – by many examples. The legal status and name of the servicing institution and the place of these services in the hierarchy within the institution are all practical, administrative questions not affecting the merit of the problem. Viewed from a scientific angle, the relationship between the two types of services can be nothing but *co-ordination.* It is, then, just as improper to look upon documentation as part of library operations as to qualify it as fully independent of the library. The former conception has long been made obsolete by practice: documentation has its own ways and methods, 'means and modes of expression' deriving exclusively from its specificity – presentation, analysis, transfer of information – which justify its independent operation wherever possible. The latter conception has never been proven, either theoretically or in practice, and as to the theoretical definition of documentation, it is *not less uncertain and vague* than that of library science. The variety of designations of information activities – 'documentation', 'scientific documentation', 'documentology', 'documentalistics', 'scientific information', 'information science', 'informatology', 'informatics', and the like are all synonyms for one and the same thing which by itself shows a theoretical uncertainty.[6] It should also be kept in mind that the concept of 'library' has also undergone changes, particularly as a result of the activities of special libraries. The information conception of libraries has made certain approaches to that of documentation. Theoretically, the solution might be found in what library work, documentation, bibliography, which is at least as 'independent' of library work as documentation, have *in common*, and in what *unites these three large spheres of information* rather than in what separates them.

Reference should be made here to *attempts at and researches in mechanized data processing and retrieval* as a recent factor strengthening the documentation-oriented approach. If the momentarily utopistic conception formulated by a Hungarian author – that 'unwritten records of mankind will be stored in a few large international centres, being not huge libraries, taken in the present sense of the word, but giant data storing machines, memories, the central register of human knowledge and culture',[7] became true, it would certainly involve revolutionary changes in the storage and diffusion of information, a change challenging the most up-to-date library work and even the value of the printed word as well as the traditional forms of publication. It would certainly involve a profound transformation of the world's cultural character.[8]

Much experience will have to be gathered to enable us to tell whether this will ever be accomplished or indeed if it is desirable at all. But one thing may be taken for sure even now: the solution of mechanized data processing is not a problem of documentation only but also one of scientific information in general, including scientific libraries.[9] Here, too, the task is to find solutions in what is common, leaving the differences out of consideration.

3. *History-oriented approach*

This approach, which perhaps should have been dealt with first for the sake of historical fidelity, appears in the clearest form, both objectively and subjectively. Obviously enough, studying historical themes (the history of the book, the history of libraries and printing) has been and will for ever be relevant to the scholarly profile of large historical libraries, inseparable from their holdings, traditions, and from the generations of librarians with classical erudition. This scholarly character has always existed and will exist as long as the traditional forms of publication survive, and even after that since the above-mentioned huge machines will never be able to substitute historical studies, codexes, old and rare books, manuscripts, and their scholarly treatment.

What the history-oriented approach represents is the *conceptual continuity* of historically developed large scientific libraries. It remains progressive as long as it assumes no aristocratic 'traits' and does not consider the study of history as the only scholarly or scientific study. But as soon as such distortions appear, other types of library (and documentation) work will be underrated and qualified – admittedly or not – as practicism, ephemeral or as 'non-scientific', etc., an attitude which will be regarded by the representatives of the affected fields as 'conservativism', 'remoteness from reality', etc., and will be reciprocated with an underestimation of the study of history. This also leads to the stiffening of views on both sides as is the case in the library-documentation dispute, although historical themes, important as they are, do not represent the whole domain of library science.

4. *Information science-oriented approach*

This approach which is the latest development of scientific information, can not yet be considered 'centred'; it is *rather tendency*, but its stormy development – particulary in the United States, but also partly in the Soviet Union – carries the promise of a new centred-approach, those of information sciences. This seems to be corroborated, among other things, by the International Conference on Training for Information Work, held in Rome, 15-19 November, 1971, at the Italian National Information Institute and by the FID. (See the Conference Edition of Papers.) On this occasion the principal items on the agenda were increasingly 'information science', 'informatique', 'computer and information science', etc.

Researches into the information sciences have emerged from three sources with the above denominations: from documentation, from 'computer science' (Computer-Wissenschaft, etc.), and from 'communication science'. It is hardly worthwhile to argue whether the *frameworks* of the theories, knowledge and practical activities deserve the name of science or whether this can be said even about the 'systems-research', individual elements of which are also integrated into information science. The substance in the application of these *scientific trends* in the field of information science is that they approach the problems of information, by relying on the highest computer techniques, in *a formalized manner and a quantified form.*

Scientific information – even in its broadest sense – is only *a part* of this complexity of problems, which points far beyond the abstracting services and the 'traditional' possibilities of other documentation services. *Modelling, formulae of higher mathematics, computer techniques* indicate the evolution process of 'information science'.

To what extent this process will be able to bring order into no longer information-flow but information-*anarchy*, how it will be capable of copying beyond the quantitative problems, with the *qualitative* ones (selection, disclosure and transmission of relevant information) – belongs to the future. It is, however, quite obvious already *today* that not even 'information science' offers us an automatic information system, but an *automated* one, where the role of the *human intellect* remains invariably decisive.

In outlining the different approaches we started from the assumption that the relative backwardness of theoretical studies is one major reason for the virtual incompatibility of these approaches and of the trends in library policy reflecting them. Virtual indeed, since theoretical investigation in library science (used as a collective term to denote the related investigations) and in science organization (showing presently and, as may be hoped, also in the future, an upward tendency) do not furnish a sound basis for the assumption that the differences between studies termed as library science (or the theory of library work), theoretical studies in documentation and historical investigations, i.e., between these disciplines, are more significant than their co-ordinatedness or their common features. These common features are definitely predominant, and the controversy over terminology, classification, and organization reflects the backwardness of theoretical studies which is accentuated by the fact that the classification and theory of science have not had much to contribute to the solution of this complex of problems. The fields discussed here (library, documentation, historical research, information science) may be considered as a dialectic unit: *all the four fields with their peculiarities form an organic part of a uniform cultural and science policy;* and this, in turn, organically includes – to use a general collective term – a uniform library system and a librarianship whose conception and interpretations also raise their particular problems.

E. THE TWO FEATURES OF SCIENTIFIC LIBRARY

Varying from one country and period to another, the type of the supervising body (or bodies) of libraries indicates the 'administrative' conception formed – or not formed – about the scientific library. Consequently, the theoretical elaboration and classification of the above-discussed problems may only permit the theoretically well-founded elaboration and further development of the conception of a general library system and librarianship.

The scientific library – including the special library and documentation, too – is an integral part of culture and of the general library system involving the various types of public libraries; it is also an integral part of the entire body of science, component of the scientific and technical revolution, and it thus has a dual aspect. This being so, the concept of the unity of library system and librarianship, taken by itself, is nothing but a fiction much in the same way as unity of education were it to appear without any differentiation of the various types of schools. Differences between the individual levels of school types and of libraries are to indicate certain qualitative differences in requirements and purposes rather than in value. In case of library types this involves a differentiation in the organization of their holdings, in their methods and services. In this sense, the scientific library is a subject of science policy and as regards planning, it should be dealt with within the given country's scientific and technological plan without, however, being omitted from the cultural plan, since it also forms a separate heading within the plan of the socio-cultural branch. Accordingly, the trend in the further development of the concept of the unity of library system and librarianship would be this: *to shift the stress towards the unity of science in the light of this dual aspect and commitment of the library.* This would be likely to have certain implications in practically every field: problems would arise in such fields as scientific qualification, scientific research, higher education and training, technical instrumentation, etc.

All this requires further investigation and also necessitates, in perspective, a many-sided, concrete inquiry into the problem of scientific libraries, and of the state of library theory, as part of the national science policy.[10] These problems are 'touchy' only if they are not made evident or are approached with impatience and mistrust without due understanding, or if questions of prestige, real or imaginary, with different values come to the fore instead of a scientific conception. However there are many important questions, theoretical and practical, which remain to be solved in the sphere of the unity of librarianship among which there is that of the relationship between general scientific and special libraries.

F. THE DIVISION OF LABOUR BETWEEN UNIVERSAL SCIENTIFIC LIBRARIES (AS GENERAL) AND SPECIAL LIBRARIES (AS PARTICULAR)

In the age of what is called the 'information deluge', the large scientific library of universal character is nothing but *a fiction* as regards the comprehensive of written documents. Carrying this problem to absurdity, such a universal library would assume all the functions now performed by the national libraries of the world, a task that cannot be tackled even by such immense institutions of 'unlimited' possibilities as the Lenin Library or the Library of Congress. And, considering from an international viewpoint, nor is these a need for such a giant institution. It may well be laid down as a principle that *'one library is no library'* since only the totality of libraries and library networks of a country – or with some exaggeration: all libraries of the world – can potentially meet all the demands of science. The document production of a country is made available to both national and international users by the country's national library, and the pooling of these documents in their entirety or even partly into one universal library is all the more unnecessary since both the content and the level of the documents are extremely heterogeneous. Universality in such a sense that a library should collect everything (even only to a defined degree) from 0 to 9 in terms of UDC is also unnecessary.

However, there is a need for the existence of 'universal' or rather general libraries whose 'universality' is taken in another sense, even within the narrower compasses of defined fields of collecting (profile), and these libraries perform indispensable functions.

These functions are the following. A special library usually covers, to the largest possible extent, literature necessary to current research projects irrespective of the trend, new fields of research, or new branches of science and scholarship appearing in the world's scientific literature. In deliberate scientific co-operation and division of labour with the special libraries, the task of the general scientific library may consist in acquiring whatever it considers of lasting value in the world literature, going into detailed acquisitions only within the scope of its main profile. *The meaning and justification of universalitv* on the one hand derive from such an outlook upon the world, and from an independence from current research projects which permits the library to follow the general development of science, and, on the other hand, derive from the fact that up-to-date scientific work is characterized by complex research assuming the co-operation of several major disciplines as well as their literature.

A scientific library may be considered to have a general profile even if it collects a selected range of outstanding works within some but not all disciplines. A further category in which the universality becomes evident is in the acquisition of encyclopedic, bibliographic, and general scientific works, handbooks, union catalogues, directories, and the like which cover all branches of science and whose centralization promotes the information of the special libraries, too. In other words: *universal scientific libraries are taking part in the division of general library work through their centralized information basis; special libraries do so through their decentralized and highly specialized holdings.* This is one of the most important forms of centralization and decentralization combined in scientific libraries. It can be outlined like this:

	General scientific library	*Special library (and documentation centre)*
Holdings	Specialized only in some disciplines; Oustanding works of the world literature; Independent of current research; Literature of new disciplines; Special collections	Detailed as required by the research (subject within one or a few disciplines and/or sectors)
Information basis	General: standard and reference works, manuals, handbooks	At sectoral level depending on the profile of the institute
Services	Nation-wide and covering international relations	Participation in institutional and sectoral cooperation, limited international relations.

It should be noted here that the conception of the unity of librarianship is closely connected with this problem since public library systems also have an important task to perform and to further develop the transmission of the services of scientific libraries.

The question of which disciplines a general scientific library is expected to cover or neglect in its following of world literature and in its acquisition policy is a practical one, to be answered for each individual case as a function of time and place. The related questions which still are to be solved include, among others, the elaboration of the feasible forms of division of labour, ensuring the necessary instrumentation, intellectual and technical, the development of the administrative network and sectoral co-operation, and – indirectly – the relationship between scientific libraries and the problem of the 'two cultures', although thematically they lead to the next chapter.

G. SCIENTIFIC LIBRARIES AND THE PROBLEMS OF THE 'TWO CULTURES'

The concept of the 'two cultures' denotes the (natural) scientific and humanistic erudition and the *Weltanschauung* (the world-view) as has been used in recent debates.[11] The reader's knowledge of this debate and of its major issues is taken for granted, it is therefore needless to outline it here. In connection with this debate, however, the question arises whether the scientific library can play any role – and if so, which role – in the formation of the 'two cultures', the two world-views and in resolving the related contradictions.

It hardly needs proof that special libraries and documentation play a considerable part in scientific and technical education, in research, and in the dissemination of scientific knowledge. Similarly, the active role of the respective sectoral libraries in the diffusion of the social sciences and humanities is also obvious. Nor is it doubtful that the scientific and technical libraries are also supposed to participate – within reasonable compasses – in propagating the knowledge incorporated in the social sciences and vice versa, while public libraries should be engaged in *diffusing both scientific fields.* What is, at most, needed here is the development of more feasible methods and forms. What is, then, the role – or more accurately the particular role – of the historically developed large general scientific libraries with sizeable historical and special collections?

In relation to the two extreme views as to the prospects of literature and the 'mechanized culture' (huge machine memories and the 'marauder machines') it can be stated that neither of them is able to provide the solution to the problem. Mathematical methods and the resulting mechanical or cybernetical methods, applications, and processes have gained ground in the fields of social sciences and humanities, e.g. in economics, demography, mathematical linguistics, machine translation, or, in our narrower field: mechanized storage and retrieval of information. This is an irresistible process promoting and enriching science which *should be furthered* with all possible means but *should not be fetishized or absolutized.* Mechanization and the machine itself are tools, means capable of increasing the capacity and effectiveness of intellectual work, without being able to substitute it (e.g. in art and literature).

Marx explained that work no longer appears embedded in the productive process, but forms a distinct activity in the course of which *man acts as the supervisor and regulator of the productive process.* As this is true for machine production, it also applies to the combination of various types of human activities and to the development of human relations.[12] Furthermore, 'it is not the direct work done by man himself, nor is it the time during which he works, but the *mastery of his own universal productive force,* the fact that he understands and takes possession of nature . . . in short, it is the development of the social individual that appears as *the pillar of production and economy.*' Hence the conclusion: 'the real wealth is nothing but the developed *productive force of all individuals.* The measure of wealth will then be not the working time but *the leisure.*' In addition to this, Marx's following statement also serves as an approach to the debate on the 'two cultures', and also to our immediate problem of the particular role of scientific libraries: 'Economy of working time is identical with the increase leisure, that is, with the increase of time necessary to the full development of the individual, which reacts as an immense productive force by itself upon the productive force of work. From the aspect of the direct process of production, this economy of working time may be looked upon as the production of fixed capital; this fixed capital is man himself . . .' 'Free time, which is partly leisure time, partly *time to realize more sublime activities, naturally changes him who possesses it into a different subject,* and his person enters into the direct process of production as a different subject.'[13]

What is, then, the particular role of the scientific library in realizing this Marxian humanistic perspective? On the one hand, it is expected to promote the process of science and production with all possible means, the possible modes of which is dealt with by a maze of library publications and a wide range of library and documentation services in order to achieve the maximum free time. This end is served, first of all, by the propagation of scientific, technical, and economic knowledge, an activity particularly emphasized throughout the world. On the other hand, the scientific library has and will have continuously to propagate the results of the social sciences and humanities, as well as the related documents and literature necessary to the full development of the individual which enrich its emotional world and develop the humane 'fixed capital'. In this sense, these two forms of the propagation of knowledge form *an organic whole,* neither of them having a priority over the other, and is only apparently a contradiction: ultimately both of them are human-centred since real wealth is nothing but 'the developed productive force of all individuals.'

However, to achieve this end it is necessary that libraries in the field of the social and humanistic studies should be brought into line with the scientific and technical libraries as regards their development. Libraries should be developed so that they might be able to meet the requirements of a time when free time will be incomparably more plentiful than now, when popular masses will pursue such *'sublime activities'* as science, literature, arts, and when not only the specialists but also the masses, relieved from the burden of narrow specialization, will study their 'prehistory' out of an interest in it, and finally, when scientific work will become a mass activity. Preparations for all this should be started now, and carried on continuously, with the organization of the holdings and with large-scale acquisition and arrangement of documents, old and rare books, manuscripts, and general scholarly works covering the individual fields of the humanistic studies whose 'technical depreciation' is negligible, if any. The handling of these documents will certainly constitute a considerable part of 'sublime activities' and will contribute to the many-sided development of the personality as opposed to works of rapid 'technical depreciation' that are indispensable today, but which will become obsolate in a few years because of the rapid development of science and technology. This applies to technical works in the first place, but, to a limited extent, to scientific works as well. Envisaging a renaissance of the humanistic studies and considering the 'two cultures' and the two world outlooks as *a whole,* scientific libraries must deliberately make preparations for all this, and have to work on the formation of this unity. It is only with this that scientific libraries might successfully play their particular role in scientific and technological development and might adequately add, not only in the scientific and technical field but also in the humanistic education, in the formation of human character and society, to the many-sided development of the individual with their particular methods.

H. SCIENTIFIC LIBRARIES AND THE 'PER CAPITA WELL-BEING' IN THE HUMAN FIELD

As has been referred to above, the tasks of scientific libraries in promoting research work and technological development, along with the related methods and forms are dealt with by a very sizeable special literature. Much attention has also been given in the literature to the general tasks, to the situation, and to the questions of developing the scientific library. It is obvious that scientific libraries have also tasks, even though indirectly or through transmissions, in the *peaceful competition* of the two great social systems of the world and they have to contribute to the solution of the immense problems of the developing countries. In the economic, scientific, and technical fields this may be formulated this way: scientific libraries can also be helpful in the effort to reach the world standard. This is precisely what sets a perspective before the library. And this is the ultimate objective of library and documentation work in these fields.

However, reaching the world standard also has its own perspective or ultimate objective which is not merely the raising of the per capita production of steel or butter, etc. to a certain level in order to attain an abundance of products which in the last resort may not be a *final end* in itself, but as has been discussed in relation to the 'two cultures', a means to develop the individual to the full; in other words, the competition is going on not only for the increase in the 'per capita' production of steel, butter, etc., but also for what might be termed as *'per capita good feeling'* or *'per capita self-consciousness'* or even *'per capita human dignity'*.

And this is by no means easier or simpler than the 'material' competition.

The interdependence between economy and consciousness is just as evident as the complexity of this interdependence and the ambiguity of interactions.

This, among other things, no one may assume that wealth by itself can automatically cause an increase in 'well-being' or can raise human dignity to a higher level or can make us happier. A man who takes his two lunches or has twice as many clothes or has twice as big an appartment, does not necessarily have twice as great a consciousness. If wealth, personal or social, were automatically to increase 'the well-being', or to imply more culture or to involve more 'sublime activities', this would certainly appear so in the historically 'luckier' and richer nations of the West. 'Alienation' and various signs of crisis in the 'well-being' do not refer to this. The reverse of what has been stated above is not true, either, at least in society, namely that in the long run *'one may dream of beauties even kneeling on peas'.*[14] Without a high level of material culture and production and without an abundance of products there is no freedom and its use in the Marxian sense, there is no possibility of eliminating narrow specialization and of the many-sided development of the personality. This might be possible for the individual or for groups with a high degree of consciousness but not for society as a whole.

It is, however, highly probable – or rather necessary – that the countries which have started their development with undeveloped productive forces and under less favourable circumstances, will yield more as regards the '*per capita* well being', human dignity or educational opportunities than the economically more developed (industrialized) countries. This is true in many respects and much has come true by now, particularly in the field of education and culture.

What we have in mind, then, is that the term 'peaceful competition' is taken to mean not only an economic competition but, considering its final objective and its possible approach, also a competition for the increase in 'well-being', both socially and individually. Following from the contemplation of this peaceful competition *as a unity of economy and consciousness* are those particular tasks of the scientific library which have already been explained in the previous paragraph. Thus, what has been stated in connection with the interdependence of the peaceful cooperation and the scientific library starts out from and returns to the debate of the 'two cultures'.

The complex participation of the scientific library in the peaceful competition and in *the elimination of backwardness* is what ultimately widens the horizons, *perspectives and social usefulness of work,* and defines the place of scientific libraries in the social division of labour in the age of the scientific and technical revolution.

CHAPTER III

An outline of the systematization of science organization

A. SOME CHARACTERISTICS OF THE SCIENTIFIC AND TECHNICAL REVOLUTION

It is generally recognized that our epoch has witnessed the unfolding of the most far-reaching scientific and technical revolution in history. The essence of this revolution and its principal elements are: automation and the complex mechanization of production, the electrification and chemization of industry, the increasing utilization of nuclear energy for peaceful purposes. The scientific-technical revolution manifests itself in the use of new kinds of raw materials, chiefly in their synthetic and technological application, in the appearance of new consumer goods, in an accelerated improvement of mechanical installations, in the development of new industries and in the rapid growth of the productivity of labour.[1]

As to the social significance and the economic impact of the scientific-technical revolution (in the capitalist system), the discussions have come to the conclusion that the wide perspectives it opens for multiplying the rate of profit permit a large-scale replacement of the old machines often three to four years after actually having been put into operation. This means that the technical depreciation of the machines has considerably quickened its traditional pace. The possibility of a wide-scale replacement of machines is one of the reasons why even during the crisis of post-war capitalism the capital investments could be maintained at a comparatively high level. The lasting economic boom in Western Europe can also be ascribed, in a large degree, to the development of the scientific-technical revolution. This has brought about such an important change in the structure of industry that the growth of production has provisionally and partially become independent of the fluctuations in individual consumption. Further, the scientific-technical changes are partly responsible for the fact that the discrepancy between the production capacities and the solvent demand has not been pushed to extremes in spite of growing exploitation. At the same time the rapid technical development and the putting into practice of the scientific achievements have also produced substantial changes in the composition of the wage-earning population (in the United States, for instance, the absolute number of the workers has decreased). Yet the number of those employed in the services has grown, and so has, in general, the ratio of those stratas of the population whose incomes are not so strictly dependent on economic expansion. This is by and large the main line of reasoning of the above discussion on the impact of the scientific-technical revolution on, and its consequences for, capitalist economy.

Even this outline survey of the consequences, the economic and the concomitant political significance of the scientific-technical revolution, shows that its effects have a bearing on the very essence of the capitalist world system. This can be summed up be saying that the scientific-technical revolution, though only temporarily and to a limited extent, introduces, nevertheless, 'corrections' into the modern capitalist reproduction cycle, and among its 'side-effects' such phenomena as the transformation of the structure of industry and employment can be observed.[2]

The relationship between social progress and the scientific-technical revolution was one of the central themes on the agenda of the East-West meeting of scientists initiated by the editors of the periodical World Marxist Review and by the Marxist Research Centre of France, organized in May 1961 at Royaumont and the Sorbonne under the title 'The future of mankind'.[3]

The basic idea underlying the discussions at this international meeting of scientists – which included such names as Academician N.N. Semyonov (Nobel Prize winner), Professor Bernal, M. Cornforth, Josue de Castro, Roger Garaudy, Alfred Sauvy, G.

Gurvitch, Jean Fourastié, etc. – was outlined by Professor Laugier as follows: Gone for ever is the time when society could afford the spontaneous intrusion of individual discoveries and inventions into human life. In our age it is becoming increasingly clear that social progress depends not so much on the scientific discoveries themselves, as rather on a rational, organized utilization of the discoveries to the benefit of mankind. Professor Bernal expressed the same idea by saying that: the main task of mankind today is to utilize in practice the new perspectives offered by science and to put a brake on the destructive force of science. And also: science – which has created new types of computing, gauging and controlling machines, the electronic machines – is acquiring a constantly increasing role in the organization of labour in the spheres of both production and management.

But the constantly increasing role of science in the organization of labour does not, in itself, solve the problems – so the reasoning goes – because our present wealth of knowledge is amply sufficient to put an end to all misery in the world. The success depends on the organization of society, on the social system. According to Professor Bernal, an average annual increase of 10 to 20% in production would enable any country to liquidate backwardness and to soar above the present highest economic standards within one generation. Again according to his opinion, the number of highly qualified manpower could be trebled within one decade, and by the liquidation of analphabetism the present intellectual resource of the world could be raised twenty-fold.

This perspective – whose economic-scientific reality was emphasized by Academician Semyonov at the Conference (who adduced data on energy resources, food production, etc.) – is, indeed, within reach; the main condition of attaining it lies with the social structure and not with the scientific-technical endowments or with the sources of labour and raw materials which are available in sufficient quantities to achieve a rapid social development all over the world.

The Programme adopted at the 22nd Congress of the Communist Party of the Soviet Union deals with the utilization of the scientific achievements and with the problems of science policy and science organization in more than one respect. One of its most important statements is that: science completely becomes a direct force in production.[4]

Two major works, published at the end of the sixties, strive to summarize and also grasp in data, the main aspects of the scientific and technical revolution. The one, under the direction of Radovan Richta, was achieved by a pluridisciplinary team under the auspices of the Czechoslovak Academy of Sciences, and the other is a work by the chief of the science policy division of the OECD, Jean-Jacques Salomon.[5] Both these works have a universal orientation. In the first, the socialist experience in this regard and the Marxist ideology are more pronounced. The second gives a comprehensive outlook about the 'politic' of the science, the inter-relations between the science and the policy and public affairs.[6]

The most comprehensive and concrete national science policy studies are published by Unesco: 'Science policy studies and documents', begun in 1965. A Europe-wide study was published in 1970 as the preparatory main document for the MINESPOL (Ministerial Conference on Science Policy) as the No. 17 of this serial 'National science policies in Europe'. The OECD published also a valuable serial on national science policies, the 'Reviews of national science policy'. Also to be mentioned, as a general introduction to the problems of science policy, is another Unesco publication: 'Le développement par la science'.[7]

In concluding on the different approaches it can be stated that science has become *a main factor* in development. Therefore, its economic, social and administrative problems need special treatment; in other words, its theoretical and its practical aspects, its management, finance, etc. require to be developed as a special area of research. This special domain is the science organization.

B. SCIENCE ORGANIZATION AS A NEW AND COMPLEX BRANCH OF RESEARCH

The appearance and development of a new branch of research and social activity, which could be called the science organization, is *a point of convergence* of the 'science of science' (a philosophical, sociological and political approach) and management (a complex activity). This reflects the consequence of the process by which science has become a force in production, or direct productive force.

1. *Science organization – as a terminological problem*

The first study theoretically formulating a new and complex discipline dealing with the problem of scientific work itself and with its social, economic and organizational implications was by the Polish sociologists Maria Ossowska and Stanislas Ossowski, already in 1936.[8]

All relevant literature on 'science of science' refers to this article as the point of departure of a conceptualization of science which takes into account its social and economical implications, as an independent sphere of research. It would be possible to elaborate a long bibliography on this matter, but J.J. Salomon in his 'Science et politique' is completely right: 'Une bibliographie complète sur les rapports entre la science et la société est, par définition, une entreprise impossible . . . tout effort en ce domaine est voué à être partiel et, pour tout dire, arbitraire. Sans doute les aspects sociaux, économiques et politiques de la science n'ont-ils fait l'objet de recherches spécifiques, hors celles des historiens proprement dits, que depuis un demi-siècle; mais le développement des activités de recherche et leurs répercussions sociales depuis à peine un quart de siècle ont à ce point accru la littérature mondiale dans ce domaine que, même sur les aspects particuliers des problèmes en jeu, il devient toujours plus difficile d'en faire le tour, à plus forte raison si l'on veut ajouter aux livres ou documents les articles de revues . . . mais, comme la face visible d'un iceberg, ils ne montrent encore que la portion la plus évidente d'un domaine de recherche et de réflexion dont l'inexploré est assurément plus vaste.'[9]

In the definition and terminology for the 'science of science', there is no common usage. This is a natural phenomenon for a new discipline. If it is considered that statistics, an activity and discipline that has a long history, has nevertheless 'more than two hundred definitions on the notion and the contents',[10] how much more is it to be expected that the 'science of science', 'science policy' or 'science organization' should still be searching for its proper terminology and definition?

As to the question of 'science organization', this term is indeed not in use in Anglo-Saxon or in French specialized literature. The parallel terms are, however, much in use in German literature ('Wissenschaftsorganisation'), and it has a large use in the socialist countries, though not in general currency. In these countries are also used: 'nauka o nauke', 'naukovedenie' (Russian), 'naukoznawstwo' (Polish), 'Wissenschaftskunde' (German) as 'science of science'. The much used terms in Western literature are 'science policy', 'research administration', 'research planning', 'organization of research'. These are all partially overlapping the field of 'Science organization', but none replaces this term. For example, the 'science policy' as notion and literature in its broader sense covers the relations between science and states but it is a definite 'corpus' only in a much larger discipline, which contains also the philosophical, the sociological and other related questions and aspects of scientific work. Such notions as 'research management' or 'research planning' specify, but only partially, a very significant expression of a much broader complex. The 'organization of science' refers directly to administrative and management aspects; it can be considered as a synonym of 'research management'.

It would be appropriate and convenient both from theoretical and practical points of view to have a *general,* a *universal* terminology covering all, or the most significant and relevant notions of the (so-called) 'science of science'. One such possibility is the adoption of the term 'science of science', but to give more adequate expression to the *practical* aspects of science, to *its economic role* in social development and in technology and so as *not to separate science from technology,* the notion 'science organization' can enter into consideration. This would be a multidisciplinary theory and activity which provide for the *amalgamation of the 'science of science', science policy and management science.* It is *in this way* that we shall use the terminology of 'science organization' and apply it for an *outline of classification* (see point C. of this chapter 'The sphere and systematization of the science organization').

2. *Some aspects and data on science organization*

The advance by great leaps in sciences and in research work can be characterized by a large amount of data. According to official sources and computations based on them, the number of the workers in research and higher education in the first two decades of our century can be put at about 50,000, of which 15,000 goes to research, all countries considered. The sum total allocated to research was half a million pound sterling.[11] In 1960, on the other hand, the number of those engaged in scientific research was some 650,000 and the research expenditures of the world can be estimated to have attained $ 20,000 million.

The first international comparison that can be regarded as official on the research input of the advanced countries related to their national income is published in a Unesco document,[12] collating the relevant data from the Soviet Union, the United States, Great Britain, France, Japan, the Federal Republic of Germany, the Netherlands and Belgium. According to the data published, the USA, the Soviet Union and Great Britain were heading the list in 1961 with 2.5 to 3.5% of their national income assigned to such purposes.

The growing – and lately steepening – rise of the share of scientific institutions, research institutes, scientific workers, and scientific budget in the advanced countries compared with any of the economic sectors proves the decisive importance of science for the development of all economic sectors (industry, agriculture, transport, communication etc.).

Science and research constitute in our age the driving power in the development of the productive forces, and within research the leading role belongs to basic research. This is confirmed by Professor Bernal with interesting – though only estimated – data.[13] He puts the number of those engaged in theoretical research as their main profession at some 80 thousand in the Soviet Union, at 40 thousand in the United States and at about

Total personnel engaged in R + D (expressed as Full-Time Equivalent - FTE) -1967

country	total R + D manpower	scientists and engineers engaged in R + D	
		total number	% of total R + D personnel
Belgium	20,957	9,010	43.0
Bulgaria	37,660	11,063	29.4
Czechoslovakia	130,874	40,734	31.1
Finland	5,477	12,109	40.0
France	184,519	50,744	27.5
Federal Republic of Germany	207,384	62,104	30.0
Greece (1966)	2,729	1,217	44.6
Hungary	31,378	10,469	33.3
Italy	49,939	19,670	39.4
Netherlands	50,200	15,700	31.3
Norway	8,063	3,512	43.6
Poland	145,903	44,978	30.8
Romania	41,775	19,231	46.0
Spain	12,988	3,486	26.8
Sweden	25,049	6,566	26.2
Yugoslavia (1966)	29,862	11,560	30.7

Distribution by sector of performance

country	total for all sectors	productive enterprise		general government sector	higher education sector
		enterprise level	branch level		
Belgium	20,957	12,003	1,710	1,312	5,932
Bulgaria	37,660	6,561	14,891	10,076	6,132
Czechoslovakia	130,874	59,280	55,646	12,482	3,466
Finland	5,477	2,528	1,169	154	1,626
France	184,519	111,400	12,195	26,574	34,350
Federal Republic of Germany	207,384	139,719	5,253	30,496	31,916
Greece	2,729	603	213	1,333	580
Hungary	31,378	3,608	15,268	8,775	3,727
Italy	49,939	29,518	2,098	8,260	10,063
Netherlands	50,200	29,700	6,250	6,050	8,200
Norway	8,063	3,009	1,153	1,615	2,286
Poland	145,903	44,200	71,577	18,706	11,420
Romania	41,775	2,639	22,673	11,492	4,917
Spain	12,988	2,962	5,110	4,561	355
Sweden	25,049	16,467	–	3,852	2,377
Yugoslavia	29,862	7,483	11,441	6,047	4,891

Funds expended on R + D in 1967 in some European countries

country	unit of national currency	total expenditure on R+D: in millions of units of the national currency	total expenditure on R+D: as a percentage of GNP	expenditure on R & D per head of population: in units of the national currency	expenditure on R & D per head of population: in U.S. dollars**
Belgium	B.Fr.	9,051.6	0.93	948.0	18.96
Bulgaria	Lev	106.4	1.36	12.8	6.40***
Czechoslovakia	Koruna	8,410.0	3.59*	588.0	38.46***
Finland	F.Kr.	194.7	0.65	41.7	13.03
France	F.Fr.	11,690.9	2.17	236.0	47.81
Federal Republic of Germany	D.M.	9,245.3	1.91	160.0	40.00
Greece	Drachma	338.7	0.17	39.3	1.31
Hungary	Forint	2,916.0	1.43*	285.0	9.50***
Italy	Lira	279,453.0	0.67	5,338.0	8.54
Netherlands	Gulden	1,860.0	2.26	148.0	40.88
Norway	N.Kr.	638.3	1.07	169.0	23.67
Poland	Zoty	10,800.0	1.79*	339.0	10.58***
Romania	Lei	1,666.3	–	86.4	5.76***
Spain	Peseta	3,636.8	0.22	113.0	1.89
Sweden	S.Kr.	1,694.0	1.37	215.0	41.56
Yugoslavia	Dinar	735.3	0.74*	37.3	2.98

* Percentage of Net Material Product (NMP) and not of Gross National Product (GNP).

** The U.S. dollar equivalents are calculated in accordance with the official rates of exchange as applied by the United Nations.

*** This figure is calculated on the basis of the rate of exchange applied to 'non-commercial' transactions.

10 thousand in Great Britain. Their total number in the whole world can be estimated at about 200 thousand, but this is rapidly growing. According to his calculations, the number of research workers in the more advanced countries shows an annual rise of 7%, i.e. a steeper rise than in any other professional branch. In the Russian-language edition of his book (1960) he contends, on the basis of computations, that by the year 2050 some 20% of the active population of the globe will be engaged in the scientific field and the number of those employed in research will exceed that of industrial workers.

From the great number of relevant official statistical data (and without any mischievous objections: it is extremely difficult to find two data or one item agreeing in the international comparisons), and only as a demonstration of the economic role of science, we shall cite some statistics abstracts from an Unesco publication, from the main preparatory documents for the Ministerial Conference on Science Policy.[14]

According to another Unesco evaluation the number of scientists and engineers engaged in the R&D in the world between 1960-1970 is estimated as follows.

country	number in thousand	% as regarding the population	% world population
Total	1,800	100	100
USSR	637	34	8
USA	497	28	6
Western Europe	264	15	9
Eastern Europe (without URSS)	173	10	3,5
Japan	115	6	3
Developing countries aprox. (Asia, Africa, Latin America)	115	6	70

In other terms, from the total of scientists and engineers in the R&D, 94% is concentrated in the areas of USA, Western Europe, Eastern Europe and Japan; these areas contain 30% of the population of the world; as for the number of all scientists and engineers this percentage is 86%.[15]

On the other hand, in books, papers and articles the conception of science as a separate industrial branch, or, as formulated by same western authors, as 'business' – and research as 'big business', since its investments are the most productive – occurs more and more frequently.[16]

More and more activities are becoming *'science-based'*.

One of the outstanding western research institutes in social sciences and economics, the Paris Institut de Science Economique Appliquée has started a series of studies on the problems of science organization. The first issue of the series, 'Problèmes économiques de la recherche et de l'information scientifique', says that research and science can be looked upon as an industry, *'research industry'*, what is more as a fundamental industry whose essential task is *to 'manufacture', to produce scientific and technological knowledge,* information. The most important *product* of this 'industry' is discovery and invention whose principal economic effects manifest themselves in *raising productivity.*[17] The series has set itself the task of analysing the 'business administration' of research and scientific information, more exactly, to contribute to its development by processing the literature on research organization. And this literature is growing at a tremendous pace, see Appendix B: Bibliography to this chapter.

When the developed, industrial countries devote 1.5-3.5% of their national income to R&D, that is to say when the R&D come to occupy a key role in economic development, it is natural that *research itself becomes an object of research.* This new kind of research activity develops its own *institutional system* and its *specialized literature.* The evolution of the scientific-technical revolution, the turning of science into direct force in production, entails the development of science organization as a quasi almost new discipline, as a new branch of knowledge and activity which, in the course of its existence, develops its own specific documentation: that of science organization.

C. SCIENCE ORGANIZATION AND ITS DOCUMENTATION THE SPHERE AND SYSTEMATIZATION OF THE SCIENCE ORGANIZATION

The theoretical and practical problems of science organization as outlined above determine by and large also the problems of documentation of science organization. Unlike the established documentation of historically established branches of science (e.g. agriculture, industry), science organization as a new branch is in a phase of rapid revolution. But this is not the only and not even the most essential feature of documentation in science organization that distinguishes it from the documentation in other branches of science. What is new, is much rather its interdisciplinary character, its evolution closely linked with all disciplines.

What distinguishes the documentation of science organization from that of other disciplines is its *universality*, its *complexity*, its character of *synthesis*.

The documentation of science organization follows up and processes the achievements of research in the different disciplines only inasmuch as these have a bearing on the universality of research, an importance for planning, organization, methodology, management or for more than one discipline. The task of the documentation in the individual disciplines is to register the specialized literature, including the problems and achievements of the research workshops (institutes, colleges, laboratories) and the activities of production units (industrial and agricultural enterprises) relating to research.

The registration of the literature on experiments revealing a new compound in a research institute or factory does not come under the heading of documentation of science organization. A publication on a *new type of co-operation* between a research institute and a production unit, on the specific forms and methods of this co-operation in connection with the experiments revealing a new compound, should obviously be registered. Another example: the registration of the literature on a new diagnostical procedure worked out experimentally at some clinic, is the task of medical documentation, but the publications on how the clinical research can be adjusted to the general medical plan or to the national research plan are to be registered by the documentation of science organization. And this is because the health spendings (scientific planning and budget) are in most cases interlinked with the training of physicians (planning and budget of higher education) and medical treatment (health service planning and budget).

Complexity is a governing principle of documentation of science organization in two senses. Firstly it imposes the task of registering bibliography on the organization and planning (but not on technicalities) of research affecting several important branches of science or considerably promoting the totality of sciences, such as space or nuclear research as well as new disciplines born from the development of sciences – it has a complexity in this sense. Secondly, it has the task to register the many-sided interrelations of research activities, including, in the first place, those of classification (according to fundamental, applied development research and technology), of methodology and organization, economics of science and scientific manpower management, as well as business administration, but not the technical and technological aspects.

The results and development level of radiation biology and the various industrial branches do not fall, in themselves, within the sphere of documentation of science organization; they are registered by the relevant specialized documentation. However, certain parts of space research as a complex, new science, which includes also radiation biology and various industrial branches, or as a research field which embraces more than one discipline, are by virtue of this an object of concern from the points of view of planning, of organization and of research organization. In the same way mathematics, affecting almost all disciplines, is not, in itself, the subject matter of documentation of science organization, yet it becomes relevant by its specific application to the various disciplines and to science organization and information (mathematical logic, operation research, automatic translation, etc.). This much as to the first aspect of its complexity. And as to its second aspect: the problems of engineering industry are the concern of technical documentation, yet its demands for qualified manpower, its research requirements, their estimation and planning as well as the relations of the development of the totality of technology and sciences, do come within the sphere of documentation in science organization.

An important feature of the complexity of documentation in science organization is that it reveals two aspects of scientific work. As a power factor: it shows, on the one hand, how science serves practice, how the achievements of research turn into material forces, how science becomes a direct force of production and, on the other hand, what and how science obtains from practice, how it is enriched through the experiences and results of organizing production and society. One of the most essential elements of documentation in science organization is to follow up and reveal these permanent interactions.

Documentation in science organization when tackling the problems of international division of scientific labour and of international relations, is governed by the principles of the universality of science and of the complexity of research. What is of interest here, is the common research in some discipline or in some scientific area covering several countries, the methods of co-ordinating research, the organizational forms of the division of labour, international co-operation and the aspects of specialization.

The dissemination of scientific information, as a constituent part of research, is associated with the documentation of science organization, not through its own technology, but through its complexity and connection with research.

While the documentation of engineering or other disciplines is, first of all, of an analytical character, helping actual research work and development, the documentation of science organization is by nature *a synthesis,* conveying comprehensive information on the problems of scientific work mainly to those controlling science policy and engaged in the management of research institutions.

'The amalgation of science and production, the appearance of new social forms and their evolution have wrought changes also in the present structure of society. The fusion of science and production, of science and the technico-organizational forms of the society introduce fundamental changes into the general structure of labour also in the field of material production, as well as in the field of management and recording.'[18]

Management in general and recording in particular contain *sui generis* the elements of classification, of systematization in the most varied fields, levels and forms of social production activities. Scientific information, which is a means of management since it is a main channel for forwarding information necessary for taking decisions, is – in this context – an *'active' recording with a specific method.* It not only registers information but *locates, selects, processes, classifies* and *imparts* it.

The classification of sources and the systematization of the apparatus 'producing' information, i.e. of the information apparatus referred to as 'active' recording, may provide theoretical and methodological bases and help to fulfill this important social function, this activity which has a practical character. All this can be referred to by the collective term of the *documentary systematization of the research apparatus.*

For an itemized enumeration of what documentation of science organization covers, see the Synoptic table, as cited below.

SYNOPTIC TABLE ON THE CLASSIFICATION OF
SCIENCE ORGANIZATION
– AN OUTLINE –

1. History and theory of science
   1.1 History of science
   1.2 Philosophy and methodology of science
   1.3 'Science of science'
   1.4 Classification of science

2. Science policy and planning
   2.1 Science and economic policy
   2.2 Science and development policy (developing countries)
   2.3 National science policy
   2.4 Science policy by enterprises
   2.5 Planning on national level
   2.6 Planning by sectors (sectorial planning: industry, agriculture, medicine, etc.)
   2.7 Planning on enterprise level
   2.8 Long-term planning
      2.81 Prognostics, projections (futurology)
      2.82 Technological forecasting
   2.9 Short-term and medium-term planning
      2.91 Planning methods and techniques applied to R&D (PERT, PATH, etc.)

3. International scientific relations
   3.1 International division of scientific labour (joint research)
   3.2 International scientific bodies and organizations
      3.21 UN and specialized agencies
      3.22 Inter-governmental organizations
      3.23 ONG (professional organizations)
   3.3 Aspects of international co-operation
      3.31 Congresses, conferences, meetings
      3.32 Study tours and the 'invisible academy'
      3.33 Exchange of publications (institutional and among scientists)

4. Administration and management of R&D
   - 4.1 On national level
   - 4.2 By sectors (sectorial management: industry, etc.)
   - 4.3 On enterprise level
   - 4.4 Legal aspects of R&D

5. Economics of science and technological progress
   - 5.1 Application of results of the R&D
     - 5.11 Discoveries, inventions
     - 5.12 Licence-policy, 'know-how'
     - 5.13 New products and process of production by R&D
     - 5.14 Marketing and trade of the R&D products and processes
   - 5.2 Budget-financing R&D
     - 5.21 By states
     - 5.22 By enterprises
     - 5.23 By foundations
   - 5.3 Evaluation of R&D
     - 5.31 By efficiency and rentability
     - 5.32 By statistics (statistics themselves and methods)
     - 5.33 By accountancy (methods)

6. Scientific manpower
   - 6.1 Education and general cultural level
     - 6.11 System of secondary education
     - 6.12 Higher education
   - 6.2 Training of scientists
     - 6.21 Post-graduate formation
   - 6.3 Academic titles and distinctions
   - 6.4 Planning and distribution of scientific manpower
     - 6.41 Planning and distribution of scientific auxiliary personnel
   - 6.5 The 'brain drain'
   - 6.6 Social and human factors affecting R&D
     - 6.61 Personnel policy
     - 6.62 Labour and social conditions, the social 'climate'
     - 6.63 Socio-psychological factors (individual and 'team-work')
     - 6.64 Monetary incentive, salary
     - 6.65 Professional and trade-union groupings
     - 6.66 Moral and political factors, social life

7. The types, levels and functional aspects of scientific work
   - 7.1 Levels of research
     - 7.11 Basic (fundamental) research, 'pure', 'academic', 'non-oriented' research
     - 7.12 Applied research (industrial research)
     - 7.13 Development (research)
   - 7.2 Complex and interdisciplinary research
   - 7.3 Research by disciplines

8. Scientific bodies and organizations
   8.1 Learned societies, academies, foundations
   8.2 Research organizations
      8.21 Enterprises (production)
      8.22 Higher education
      8.23 Research institutes (independent)
   8.3 Non-research organizations
      8.31 Offices (civil service)
      8.32 Cultural-scientific institutions (libraries, museums, archives, etc.)

9. Methodology, techniques and equipment of scientific work (Material and intellectual equipment and the infrastructure of scientific work)
   9.1 The equiping of, and experimentation in R&D
      9.11 'Big' equipments
      9.12 'Little' mechanization (g.e. punched cards)
   9.2 Concrete research methods and techniques
      9.21 Mathematical methods (g.e. operation research)
      9.22 'Computerized' techniques
      9.23 Other methods (e.g. interview by sociological methods)
   9.3 Scientific information
      9.31 The theory of scientific communication, including the theoretical problems of scientific librarianship, bibliography and documentation
      9.32 Dissemination of scientific information, including mechanized information storage and retrieval

1. *History and theory of science*

Under this heading we deal with the theoretical and general problems of science.

Works on *science history* (also the vast historical surveys of the individual sciences) are instructive also for science organization, science history being, at the same time, the *history of science organization,* and suggest conclusions also for science policy. In many respects the history of science is also the history of the development of the productive forces, contributing thereby to understanding the process in the course of which science is becoming a productive force. As Marx put it: The first thing we need for investigating the relationship between intellectual and material productions is to conceive material production not as a general category but in a definite historic form. Thus, for instance, the type of intellectual production corresponding to the capitalist mode of production is different from that corresponding to the medieval mode of production. If we fail to conceive material production in its specific historic form it is impossible to conceive what is determined in the corresponding intellectual production and the interaction of the two and: a definite form of material production procedes, first, a definite stratification of the society, second, a definite relation of people to production. The structure of their state and of their minds are determined by these two. And so is also the nature of their intellectual production.[19] The works on the history of science are also important parts of the cultural history of the world and of individual countries, including the history of great inventions, expeditions, studies and the biographies of the great scholars.

The works on the *philosophy of science* analyse the motive forces promoting the development of sciences, shed light on the relationship between social progress and the sciences, elucidate the impact of the social system on science (and vice versa), the place and role of the sciences in the society and their interrelations. The works of this category, together with the history of sciences, analyse the struggle between the progressive and reactionary forces in the sciences, the struggle between materialism and idealism going on since the birth of science. This category includes writings on the ideological problems of the individual sciences, surveys of principle of the development trends in the various disciplines and, in general, works on the role of dialectic materialism in scientific activity.

The *classification* problems of science organization (including those of terminology) could not be separated from those of science theory. Parallel to the rapid rate of the development of sciences, to the growing extent of their economic and technical application, new branches of sciences have appeared and their interrelation has become more and more complicated since the second half of the past century. It is significant that already the second half of the 19th century books were published which can be intitulated by our contemporary terminology as 'science organization' publications, and it is no less significant that these books have been reprinted recently due to the increasing interest in the 'science of science' problems and in its history.[20] A specific scientific methodology and technology have developed, and scientific work has become collective in our century. Research has become part and parcel of the division of labour which has made its appearance within the sphere of research too. In the course of this development, problems of the philosophy of science and classification have been and are being raised at an increasing rate. Socialist science theory and western science theory, for instance, differ, to some extent, on what is to be understood by the social sciences.

The importance of the problems of classification is well illustrated in the Auger Report, commissioned by Unesco and UN[21] on the peaceful utilization and co-ordination of the achievements in scientific research. One of its main tasks is to classify sciences (the natural sciences) and to systemize research work (levels, directions, organization fields etc.). In science organization systematization has a considerable practical significance (trends in the development of research, higher education and the training of scientists, the sums to be allocated for these purposes, the utilization of the results of research, the planning of complex research etc.). In this respect it still remains to be clarified theoretically what exactly belongs to certain spheres of research (e.g. basic and applied research) and of development. These problems are developed also from the *social science aspects* by the one other main Unesco report.[22]

The works treating *'science of science'* as a new discipline, are also an organic part of this first category.

2. *Science policy and planning*

Belonging to this category are all indications, measures, decisions providing for the development and co-ordination of the totality of scientific work within a country, its harmonious and rational adaptation to society and state life, its co-ordination with the different fields of productive labour.

In this respect in the socialist countries scientific work, like all other social activities, is part of the national economic plan and its direction and conditions are determined and ensured by the state, while in the capitalist countries the state is only partly responsible for scientific work: a substantial part of research is the monopoly of the monopolies. This, naturally, does not prevent such research from being highly successful or even socially useful. Along with the strengthening of the trend of state monopolistic capitalism and with the growing share of military expenditure, the role of the capitalist state in science policy is also gaining importance. In the western countries a significant part is played by foundations financing research (Rockefeller, Ford, Carnegie, Volkswagen, Agnelli, etc.) whose creation and operation are linked also with the taxation system (allowances and exemptions).

In the socialist countries science policy, like economic policy and all important manifestations of social life, are directed by the central committees of the communist parties. Among the most important documents of science policy are the relevant party and government decisions, laws and decrees, the national economic plans and the state budgets.

In the western countries, in addition to state budgets, resolutions and decisions of the authorities responsible for science policy, the reports (business reports) of the big enterprises and foundations, reflecting science policy, deserve special attention.

The *planning of research* is performed on three levels: national level, by sectors (industry, agriculture, etc.), and on .enterprise level. Research is influenced in two aspects by production: firstly, by production as the result of the uninterrupted and expending development *of* research, and, secondly, by production as a process constantly imparting experience, impulses and achievements *to* research. These two aspects of production and research determine the increasingly active social character of science.

The planning of scientific work includes – among other things – the assessment of the national research capacity, its co-ordination with the research trends in the national economy, including the social-cultural-health sector and themes of a theoretical character affecting all or the majority of the disciplines. The national research plan, in the form of short- and long-range planning, is drafted with due regard to these factors – and other factors, for instance, international division of labour, availability of cadres – and to the financial capacities. With a view to implementing the aims of science policy, this plan comprises the determination of the *trends* and *proportions* of research, the priority of the most important research themes and their co-ordination. The *scientific thematic planning* must be co-ordinated with the *economic planning,* with the investment, personnel and material budget and funds for research. This category includes also the methodology of planning the methods of projections, of forecasting and the *techniques* applied.

When establishing the organizational framework of scientific work, what and where to study most efficiently, many factors should be considered: the possibility of concentrating material means and manpower; attachment to production as a direct experimental basis for manufacturing prototypes; the links between certain theoretical research and higher education; the organizing and financing activities of scientific organizations and institutions; the facilities of developing contemporary large-scale research within the traditional and developed scientific organizational frames, etc.

The documents of science policy contain surveys of the scientific situation in the country, its connections with other social activities, the directions and proportions of research, the development plans; hence, owing to its radius of action to its validity for all branches of science, the literature belonging to this category has its place in the immediate vicinity of the theoretical literature.

3. *International scientific relations*

The fostering of international scientific relations, the development of the international division of labour and the organization of research embracing several countries are among the tasks of science policy.

Joint research has entailed qualitative changes in the international scientific relations. Beside the traditional forms of these relations (such as the organization of international scientific conferences, study tours abroad, the international exchange of papers and publications, the development of subjects for joint research) international scientific co-operation and specialization in content enable the participating countries to engage in such research, and in such a degree, for which they are best equiped and possess the best endowments. Joint research also permits a country to take part in research that would go beyond its material and scientific capacities.

The socialist countries co-ordinate their economic development plans within the Council of Mutual Economic Aid, creating and evolving thereby their scientific and technical co-operation as well.

In the Western countries the state has only partial control over the research capacity of the country but the tendency towards international co-operation is accelerating because of the need to centralize the intellectual and material resources (e.g. the science division of the OECD). Parallel with this development the research activities of the monopolies also indicate a tendency towards integration (IBM, Philips, Siemens etc.).

Within the UN system and the regional intergovernmental organizations or the non-governmental international scientific organizations a co-operation has been developed on a world-wide scale (e.g. IAEA, WHO, WMO, Unesco, ICSU etc.).[23]

Research programmes on a world scale of the type carried out, for example, by Unesco in geophysics or in hydrology, would be unimaginable without international co-operation. The same, however, is true in respect of certain important social science topics, such as the study on an international scale undertaken by Unesco's Vienna Social Science Centre on 'time budget' (how people spend their free time).

In 1961 the Research Organization Unit was established by Unesco to study the problems of science organization, which then became the Science Policy Division.

A significant part of the literature belonging to this category is devoted to the problems of scientific and cultural development in the *developing* countries (technical assistance etc.).

The problems of international co-operation of scientific labour are part and parcel of science policy, and in connection with the international division of labour, the documents on the operation of international scientific and research organizations evolved in the course of recent developments are deserving special attention.

4. *Administration and management of R&D*

While international co-operation constitutes the 'foreign policy' part of science policy, the administration and management of scientific work are directed to the actual application of science policy, constituting thus its 'home politics'. The two are linked with organic, permanent ties.

Organization and administration are performed essentially on three levels: on that of state administration and for national economy, on the level of industrial branches, and on the enterprise level (in the research institutes). Consequently all three levels have their own specific problems and methods. But neither the tasks to be solved nor the methods applied can be attributed exclusively to one or the other of these three levels because they are in interaction, are organically linked and interdependent in much the same manner as the organization and administration of research are.

The organization and administration of scientific work must provide in a factual manner, and permanently, for the organizational, control, staff and material conditions of implementing the scientific plan. The management must keep an eye on the practical aspects of co-operation and specialization, of the co-ordination of research. It should ensure the practical application of the scientific achievements in the national economy, the quick and efficient diffusion of new results and the utilization of research. It should also provide for the most rational organization possible of the research institutions.

The tasks of the organization and management of scientific work outlined above affect also the content of research achievements because up-to-date research has a collective 'large-seal' character, and its success depends not only on the genius of a certain scientist or scholar or on the standard of the research workers but also on organizational work, and that to a great extent. The standard of researchers is again affected by planning and organization (training of scientific labour, distribution, research conditions, exchange of experience, documentation, facilities of publication, etc.). The research workers engaged in individual and isolated studies are being replaced by large research teams aided by extremely complicated and costly equipment and installations, capable of conducting research affecting several disciplines and requiring their collaboration. The operation of such a complex and complicated apparatus, the building up of its mechanism constitute organizational tasks requiring themselves sound scientific informations.

Consequently, the documentation of the relevant experiences, of the management theory, administrative procedures and methods is one of the important fields in science organization.

5. *Economics of science and technical progress*

In accordance with the economic significance of research, the economic problems of scientific work have assumed greater proportions in two senses. Firstly, research is a social activity which has far-reaching economic consequences affecting all branches of production and which produces values itself. Secondly, research requiring finance maintenance, operation and development raises certain economic problems.

The major topics within the sphere of the economics of science are:

Discoveries and inventions, their realization, utilization, introduction and application in production, their patenting, the system of patents and innovations; the transfer of technology (including the problems of 'technological gap') and the concomitant technical development; the scientific budget (related to the state budget, its share in the national income); cost norms according to the size of research enterprises (economic efficiency); the elaboration of computation methods and systems of indicators for each research level showing the economic efficiency of research; the efficiency of scientific investments; science statistics; the system of scientifically and economically controlling and registering research activities; the accountancy system of the specific features of scientific work.

This enumeration is far from exhausting the economic problems of scientific work, it sumply indicates their content.

Beside the priority of determining the subject of research, the economics and the economic planning of research is acquiring a growing significance, particulary in areas where research is *directly related to production,* or where research or laboratory work is identical with production (i.e. isotope production).

A new branch in economics and in national economic planning (in addition to industrial, agricultural, financial etc. economics and planning) is in the making. These are *research economics* and planning.

This process acquires intensity to an extent that the ratio of science in the national income and in the budget increases.

6. *Scientific manpower*

One of the key problems of the productivity of scientific work is the planning of manpower, including the training of new scientists. This is a long-range planning problem since the training period of an independent research worker is approximately ten years (including the period of higher education), varying according to scientific disciplines. Basic scientific training considered in a broader framework starts in the public education system. Accordingly, scientific manpower economy and training includes the whole educational process (for instance, mathematics and languages in the secondary schools) and, in certain respects, also adult education (propagation and diffusion of scientific knowledge).

Thus the sphere of scientific manpower and staff training involves the following major problems:

The system of public education, and what is important from the point of view of scientific development; the system of secondary training in the natural and technical sciences; higher education, the system of college and university training, its connections with practice, with production work; syllabuses and curricula corresponding to the demands of the national economy and of scientific research; training of scientists; the scientific practice of college graduates, organized forms of training; scientific degrees and scientific qualifications; the distribution of scientific workers, planned distribution according to the demands of the national economy and research over research levels and disciplines; material interests, remuneration of scientific work, awarding of bonuses, various forms of monetary incentives; socio-psychological factors affecting R&D.

The scientific manpower planning is closely linked with the long-range plans of scientific research and national economy and with demography (the division of the population according to professions, to economic sectors, the composition of the active population according to age groups etc.). In compliance with the growing importance of natural and technical sciences, the training of engineers and technicians has come to the fore, i.e. the training of experts in fundamental research and in such sciences as have a bearing on all or most branches (biology, mathematics, cybernetics etc.). Such aspects of expert training as relate to the complex character of contemporary production processes and research (chemical processing, automation, complex mechanizing) or as require multilateral training on account of the high degree of organization of production and research (for instance, the training of economist-engineers), have also come to the fore.

Scientific manpower planning, as one of the essential elements of the long-range general manpower planning for research and national economy, is – on account of various subjective factors – a most intricate methodological task. A much debated subject in the literature on science organization of the past ten decades has been the scientific manpower problem and, since international comparisons are of great importance in this field – i.e. the number of highly qualified specialists is an indicator of the peaceful competition of the two social systems – a large amount of statistical data have accumulated. Their collecting and processing and rendering them measurable and comparable to a certain extent constitute also an important task of documentation of science organization.

7. *Types, research levels and functional aspects of scientific work*

The problems of planning, organizing and administering scientific work can be encountered on all levels of research work, which can be assigned to three main groups accepted internationally (though the terminology cannot be regarded as finally settled, nor is the content of the groups definitely determined): basic, applied and development research.

There are no strict boundaries between these three levels of research, and it has often happened in the history of sciences that a task tackled as a problem of applied or development research resulted – during its solution or as its 'by-product', or again owing to the changed condition during research – in an achievement of a basic-research character. And the other way round, theoretical research work has also produced – adventitiously or as its 'by-product' – a result of an applied-research or technological character.

The general aim of a basic research is to detect new laws which are applied in practice and made use of indirectly. Basic research has a determining character for scientific development as a whole. Applied research is meant to make use of the newly detected laws in science and production, while development research serves direct production targets of a technological character.

The definition of these notions – and in general of research categories, according to their contents – their correlation and systematization are problems and subjects constantly discussed by science-political authorities and in the literature.

8. *Scientific bodies and organizations*

The creative workshops of research can be divided into the following major fields:

Within production or activities associated with them; within higher education; within research institutes; and within non-professional research organizations where research activities are of an auxiliary character, e.g. scientific and research activities conducted by official bodies, such as statistical standardization offices, and such scientific organizations as libraries, archives, museums.

Research work in all these fields has its specific organizational forms, frames which depend on the directions in research (natural sciences, engineering, social sciences) and on its relations to the scientific branches and the national economy. The registration and analysis of the nature of these relations and contacts in order to ensure rationality of research organization, constitutes one of the important tasks in science organization. Such questions as to what organizational frame fits best certain types of research and what are the most efficient organizational forms, are important. These experiences will be documented in the literature concerning the operation of research organizations: reports about annual meetings, annual reports, yearbooks, operation descriptions, statutes, directories of scientific organizations and similar work of reference.

9. *Methodology, techniques and equipment of scientific work (material and intellectual equipment, and the infrastructure of scientific work)*

Before the development of up-to-date large-scale research, the methodology of research work could be summed up comparatively easily since it was restricted, by and large, to the description of the specific experimental methods of the individual scientific branches, to the methods of collecting and processing data and materials. This kind of literature has not lost its importance since.

But 'research industry' can only operate on a large-scale basis, with a large apparatus, with expensive equipment and installations for experiments, and if well provided with instruments. A literature on these material conditions of research, on the new methods based on mathematics (cybernetics, operation research etc.) and on other mehods, has grown considerably in the past decades.

This literature is, at the same time, dependent on the individual disciplines because – beside certain general questions associated with scientific work, such as the organization and hygiene of intellectual work, the methods and techniques of collecting data, the processing of literary sources etc. – the two large scientific spheres (the natural and technical sciences and the social sciences) and their different branches have their own specific methodological problems deriving from the subject of research. These two large spheres of sciences are in a constant productive interaction with each other. We are witnessing the 'mathemathicization of the sciences', the endeavours to quantify the phenomena of the social sciences.

Mathematical methods have been generally adopted in economic sciences (in the socialist countries this process began in the last few years, as a result of the demand for more exact planning methods in the national economy), the application of electronic machines have widely introduced mathematical logic and statistics – that is methods based on them – into linguistics. These methods also contribute to the evolution of complex interdisciplinary research being one of the typical features of contemporary scientific work. And one of the major problems of complex research is organization, hence also the methods of organization.

The documentation of science organization is concerned with the research methodology of the individual disciplines as far as they have general (or generalizable) characteristics, and, in the last analysis, with the methods and procedures of organization and mechanization that can be applied generally to research work.

The problems of *scientific information* are part of the documentation of science organization only in so far as they are concerned with research, and not by virtue of its general 'technological' relevance, which, lately, has been the concern of libraries and documentation.

The problems of scientific information should be considered by the documentation of science organization under two aspects:

The *theory of scientific communication,* including the theoretical problems of scientific librarianship, bibliography and documentation, the analysis of publication types, structural problems of scientific publication (abstracting, indexing, apparatus of editing etc.), the analysis and assessment of original, primary scientific publications from the angle of their utilization (references, notes to publications, their mathematical and other assessment methods); and the application of new forms of publications in research (i.e. microreproductions).

To the questions of the dissemination of scientific information including mechanized information storage and retrieval, belong the application of automatized systems, some theoretical problems of bibliography and classification, the analysis of the variants and economico-scientific efficiency of documentation work, the assessment of the direction of scientific information toward centralization or decentralization in the different branches of science, etc.

Scientific information in this respect in its entirety is a category of science organization, as illustrates the following:

documentation of science organization → scientific information → science organization

or:

science organization → scientific information → documentation of science organization.

D. THE CHARACTER OF THE DOCUMENTATION OF SCIENCE ORGANIZATION

The sphere of the subject of the documentation of science organization does not fit into any of the existing science classification, library or documentation systems. Just as science organization itself appears as a new branch, as a new discipline, its documentation is developing into a new branch of scientific information which has its own specific problems of terminology, subjects, methodology and classification.

The documentation of science organization covers, not only the influence on the society, on technical development, on practice of science in general and of research in particular, but also the effect of production constantly exerted on science, on research. Not only the transfer of scientific achievements into practice but also the theoretical generalization and further development of practical results (e.g. technological change) belong to science organization and to its documentation.

In such documentation the modes of selection and registration acquire added importance on account of the complexity of the subjects: decisions of government authorities, laws, decrees, budgets, reports of annual meetings of institutions, institute reports, scientific yearbooks, monographs, papers, etc. Preference should be given to revealing the central ideas, theses, new approaches to problems, attempts to solve questions, possible modes of solution and to international analogues. The documents should be analysed for their theoretical, classification, administration, methodological and economic aspects. The accessibility of the descriptive parts, the publication of data and materials: plan reports, plans, budget data, statistics, descriptions of organizational structures, etc., are also tasks of importance. The documentation of science organization obtains its informations from heterogeneous material.

In compliance with the characteristics deriving from the complexity of the documentation of science organization, the methods of registration should approximate more and more the creative methods, more exactly, their phases of data-collecting and -processing. In addition to offering titles, abstracts and other usual bibliographical and documentary data, they are meant to convey the contents of the documents to the research workers, without, however, blurring the line between research and scientific information.

Beside acting as a transmission agent between specialized literature and research work, this type of documentation may become *a specific transition between documentation and research work.* To use an analogy taken from industry, its task is not simply to 'provide' research with 'material', to meet its 'demand for raw materials', but also to turn out 'semi-finishes research products'. In general, these are the perspectives of development for scientific information work. Drawing conclusions, evaluating, qualifying and generalizing the value of the data and attaining scientific generalizations, is creative work to which the documentation contributes.

By *'semi-finished research products'* we mean the processing of documents into tabulation, the rendering of the data methodologically commensurable (for instance the statistical data of the different countries), the evaluation of their proportions, the description of documents (organizational and administrational schemes, internal structure of research institutes, scientific budgetary systems, computation methods for assessing economic efficiency, etc.) and their international collation. By fulfilling these tasks the documentation of science organization performs also tasks of a *limited* research character, However, it does not analyse science policy directly on the basis of these data, does not generalize or use hypotheses, does not conduct experiments – these being already the tasks of research.

In addition to the possibility of generalizing from the theoretical and organization-methodological problems, the documentation of science organization has its specific organizational characteristics and forms, deriving from the concrete needs of national science policies. These organizational frames are in a constant transformation all over the world depending on the place and role of science.

Thus it seems appropriate to centralize organizationally the documentation of science organization, in certain questions, such as the registration of reports on study tours abroad undertaken in the field of R&D, extending to them the copyright deposit, a national registration of research subjects, and the copyright deposit of studies finished but not yet published. The centralization should cover small but essential tasks having an economic bearing, such as the indexing of scientific translations in the country, the copyright deposit of such translations, and the organization of similar indexes which may efficiently contribute to science organization.

Science organization and consequently its documentation directly follow from the process of science becoming a direct force in production. This process may be illustrated as follows:

scientific-technical revolution → science as a direct force in production → science organization → documentation of science organization.

This shows the last part of this process as the organic completion of its first three elements. Since we are dealing with a rapidly developing process, it necessarily contains debatable and problematic elements with respect to the subject of the documentation of science organization, to its systematization and to its relation to sectional documentations.

The centralized documentation of science organization is interconnected with sectoral documentations in institutes and plants. These latter functions are on a more restricted area of research. The sectional documentations are gradually being completed with the subjects of the documentation of science organization as far as the pertinent disciplines are concerned. This makes it possible to achieve a methodological collation and exchange of experience in science organization between disciplines, institutes, plants, domestic and foreign activities and completes the national scientific information network covering all scientific activities in this direction.

Thus, the documentation of science organization becomes an element of scientific work and of economic development on the basis of uniform principles and methodology, by combining centralization and decentralization, co-operation and specialization, to a certain degree itself turning into an economic category. And this, at the same time, determines its place in the social division of labour.

A diagram in broad outlines of the process of research and scientific information

RESEARCH	COLLECTION AND RECORDING OF INFORMATION	SCIENTIFIC INFORMATION
GENERAL EVALUATION OF DATA		ACTIVE RECORDING OF INFORMATION
CLASSIFICATION OF INFORMATION AND DATA		LOCATING
HYPOTHESIS		SELECTING
EXPERIMENTATION		PROCESSING AND CLASSIFICATION
CONCLUSIONS		STORING RETRIEVING
SCIENTIFIC GENERALIZATIONS		TRANSMITTING AND PROPAGATION

CHAPTER IV

The points of divergence between research and scientific information

Contrast, paradox, tension, discrepancy

'The third equal partner . . .'

As has been pointed out by J.D. Bernal in his *Science in history*, the appearance of science as an important social force is a decisive and unalterable step. This decisive and unalterable process manifests itself in series of divergences both within science and between science and other social spheres. To consider these divergences, to explore and eliminate them is essential for the development of science. Accordig to R.L. Ackoff, the word 'science' has two meanings: first, it means an *activity*, i.e. an organized process of research, and it also denotes the result of this process, i.e. a certain *knowledge*.

One category of the divergences existing in the present structure and development of science is formed by divergences *between science and scientific information*. These divergences are connected with the first aspect of the double meaning of science, i.e. with 'science' as an *organized process of research.*[1]

Contemporary science has reached a level where the efficiency, productivity, and performance of research workers increasingly depends in good organization, adequate information, and on the operation of the *system of scientific information*. From the intrinsic laws of the development of science and the demands of social practice has evolved – as a result of the social division of labour – scientific information, *a special kind* of scientific work.[2]

One internationally known expert of the physics of elementary particles approached the problem from a different angle but came to the same conclusion, when speaking of the *dynamism of culture* formulated this process as follows: 'The mass of information amassed by the specialists and scientists . . . has to be processed and the essence needs to be extracted from it so that no real value might be lost and be suitable for a synthesis . . . Thus a new group joins forces with the specialists working at laboratories and with those who, trying to imagine the inimaginable . . . this group is the community of *specialists of scientific information.*'[3]

The appearance of this 'third equal partner' in science is an objective phenomenon, and the concomitant divergences are also as real.

'The number of workers necessary to make a table, the amount of work of a definite kind necessary to manufacture a given product under certain production conditions are exactly known. This, however, is not the case with many 'non-material products'. The amount of work required to achieve a certain result is just as approximative as is the result itself.' This is how Marx determines the main characteristics of 'non-material production'.[4] As a matter of fact, scientific work or research is a dialectical process mainly characterized by the fact that its outcome can only be estimated and its original aim is subject to changes and modifications during actual research until it assumes its final shape. This is due, among other things, to the fact that 'the product is inseparable from the act of production, and this is the case, for instance, with all performing artists, orators, actors, teachers, physicians, priests etc.'[5]

The formulation of the research theme is connected with scientific information which must indicate whether the planned research has been carried out or initiated elsewhere. When it is made clear that the subject in view has not yet been tackled, an investigation of the literature precedes the formulation of the research task. Once this is defined, the actual research process will involve various forms of literary research and information. The first results are then published in papers which – through the medium of scientific information – are fed into the 'information fund', and the cycle begins again. At the same time pieces of information that are 'non-productive' or redundant or seem to be so at a given time also enter the cycle as partial or negative results. When talking of sciences, Descartes says that one should bend one's mind to study even those which seem most absurd and false ('Penchez-vous . . . même sur celles qui paraissent les plus absurdes et les plus fausses'), in other words, the sources of knowledge cannot be determined in advance. All the above aspects of research carry the germs of the points of divergence existing between research needs and scientific information.

A. TYPES OF DIVERGENCE BETWEEN RESEARCH AND SCIENTIFIC INFORMATION

1. *The divergence between the definite research demand and the relevancy of an indefinite body of knowledge*

All kinds of research, irrespective of its level and field of application, even what is called 'pure research' or research without any practical purpose, i.e. deriving from mere 'curiosity' rely in some shape or form on an accumulated body of knowledge, on our 'collective memory', which finds its expression in books, periodicals, and other documents. Even the most abstract scientific activity expresses a definite research demand in this respect. But in the course of most concrete development research, this research demand cannot rely on an exactly defined and specified area of knowledge or on its exploration because most of the sources from which the necessary knowledge can be gained are unknown factors. Only the area of the potential sources is known. Of this were not so, research could be regarded as solved for the most part, or no demand for research would arise. It is a different task to ascertain whether or not a scientific problem has been solved or whether the research theme is an entirely new one. At any rate the task of acquiring the necessary answer from an unstructured mass of knowledge (the literature) to meet the definite research demand appears in scientific information as a contradiction.

2. *The contrast between the immensity of the body of accumulated knowledge (general) and its relatively small 'productive' exploitable part (specific)*

There is a disproportion between the total mass of accumulated knowledge and the part that actually can be utilized by the given research theme. The *World Bibliography of Bibliographies* registers more than 84,000 independent bibliographies,[6] the Central Information Department of the Saltykov-Shchedrin Library, Leningrad, has about 90,000 volumes of bibliographic and reference publications in its special information collection.

According to a Unesco publication, experience shows that, in spite of the specialization of the periodicals, a research worker finds only five out of 100 papers which he can use in his research.[7] There is no shortage in information sources but the question arises whether and if so, how, the necessary material can be selected from a wealth of several millions of items registered in 84,000 of special bibliographies, and whether the really 'productive' literature necessary for research can be found among the annual average of 7,000 items contained in the Unesco bibliography of economics. According to some opinions research must rely on experience and on intuition, for otherwise it is hopeless to pick out the pearl oyster from the sea of publications. Indeed, there is some truth in Baudelaire's simile that the mythical albatross is hampered in its flight by its wings which have grown immense. The vast amount of information not only aids but also impedes research.

3. *The paradox that at the same time that there is the exponential increase in the number of scientific works, there is the exponential 'technical depreciation' of the knowledge it contains*

Scientific products, including publications, redouble every ten to fifteen years according to international estimates. At the same time, it is an empirical fact confirmed by various computations that the obsolescence, the 'technical depreciation' of our knowledge is also exponential. Obviously, there are differences in durability among the various research levels and between technical and social sciences. A linguistic work published a century ago may have a source value right up to the present date, but a technological or agronomical publication of the same age is of interest only for the history of science, and even a technological work of a few years ago may have become obsolete. Some special libraries have as their policy the periodic replacement of material no longer appropriate.

According to the computations of British experts, the interest in publications in natural sciences and technology starts to decline at a rapid rate after one year and the publications become, in general, entirely obsolete after ten years.[8] According to other assessments made by Czechoslovak specialists, 24 to 28% of the publications outlive a decade, and 12 to 15% of them still recur in references within twenty years but then become redundant.[9]

As far as information is concerned, this process of redundancy is also complicated by the fact that only part of the content becomes outdated:[10] the results published are extremely short-lived whereas the methods applied, certain data and tabulations, and – more important – certain descriptions or laboratory results that might have seen adventitious may nourish thought or turn out to be valuable contributions in some context other than the subject of the publication.

The fact it is not possible to assess exactly the rate of 'technical depreciation' of scientific information and that the pertinent assessments show some discrepancies does not alter the picture as a whole, i.e. the exponential character of this depreciation. Yet in connection with the increase of the number of scientific products we call attention to Engels's statement who, arguing against Malthus's conception, says: '. . . science and its progress is just as infinite and at least as rapid as that of population . . . Science expands at least as rapidly as population, the latter growing at the rate of the last generation; science advances in relation to the mass of knowledge handed down by the preceding generation, that is, under the most common circumstances, in a geometrical ratio . . .'[11]

4. *The paradox that at the same time there is differentiation and integration in scientific literature*

This is a dialectic process reflecting the development of sciences in the interdisciplinary fields and their tendency to specialization. In technical literature, especially in periodicals, this paradox can be noticed in the extraordinary widening in the coverage of the publications and of the papers themselves, while individual themes become narrower, more specialized and in the continuous schizogenesis of the publications.

According to Engels, the process of specialization and differentiation in sciences started as early as the end of the eighteenth century: the eighteenth century summed up the results achieved in history until then . . . explained their necessity and their internal relations . . . the sciences gradually improved while being attached to philosophy, on the one hand, and to practice, on the other.[12] Engels also raises the question of the integration process which he designs in connection with the activities of the French Encyclopaedists, as the expression of the demand for recognizing the relationships between things. These two interdependent processes, or more correctly these two sides of one process – that of the development of sciences and that of its reflection in the literature – attain their full evolution in the age of the scientific-technical revolution when not only entire specialized disciplines develop within the natural and technical sciences (e.g. biometrics, bio-meteorology) and in social sciences (e.g. industrial statistics, planometry, sectoral economics) and also on the borders of these two wide ranges of sciences, but the results and research methods of these two vast domains penetrate into one another (for instance, the application of cybernetics in the most varied fields or that of management science in various branches of sciences and practical activities).

At the same time branches of knowledge, scientific activities, affecting all sciences, begin to develop, the definition and location of which within the system of sciences take some time. Cybernetics or operation research themselves, for instance, have no unanimous, exact, and generally adopted definition for the time being. Branches of knowledge of a general character, namely management science, labour economics, science organization, scientific information, can all be assigned to the social sciences yet their action radius covers the natural and technical science researches too. A typical example is given by labour economics which has acquired independence during the process of specialization of social sciences and turned into a discipline integrating the elements of several other branches. The evolution and application of management science, of science organization, and of those branches of planning, organization, and administration which are using a mathematical apparatus such as operation research, linear programming etc. display a tendency in the evolution of social, economic and scientific life of emphasizing management and organization.

All these intricate processes in the presentation of specialized literature appear as contradictory in scientific information, which itself takes part in the scientific foundation of organization and management.

5. *The tension between the social demand for the increasingly wide application of scientific knowledge and the increasing complexity of that knowledge (by scientific knowledge, we understand the natural and the social sciences)*

Scientific studies and research are becoming more complicated and exacting both in general and in the field of social sciences on account of the accumulation of literature (world-literary character) and of the development in the interdisciplinary fields. Even if compared with the situation some twenty years ago, the research in social sciences must cope with a much larger written material; at the same time the use of mathematical methods presupposes the mastering of a new body of knowledge. New research areas develop, and under the impact of these factors the study of sciences becomes more and more sophisticated. But at the same time, as a result of the scientific and technological revolution, work as a whole becomes intellectually more exacting, and requires the mastering of ever more scientific information. In the socialist countries one of the principal aims is to reduce and then to liquidate the differences between intellectual and physical work, which presupposes not only a high-level and massive automation but also a large-scale widening of the technological and social-scientific knowledge of the largest part of population as possible. The idea is not only to train the greatest possible number of research workers but also to develop the scientific nature of knowledge at all levels of work, i.e. the intellectualization of labour, which, owing to the growth of complexity of the sciences is not a linear process.

But the world-literary character of modern science entails not only quantitative problems but also linguistic ones. In the nineteenth century owing to the variety of periodicals, the knowledge of one or two major languages was sufficient to keep up to date. According to an assessment, within every major professional field there were 32 journals before the Second World War II, 14 before the First and 2 in 1900 as against the 50 in our days,[13] in the second half of the twentieth century the growth and expansion of the scientific literature has raised considerable linguistic problems all over the world. The cessation of the scientific hegemony of the English, French, and German speaking areas, the achievements of soviet science (space research, mathematics, physics etc.) and the advancement of scientific life in the people's democratic countries have raised new aspects to the language problem. The European translation centre for example has as one of its main functions to keep record of the specialized literature published in the Eastern European countries etc.[14] And in the socialist countries the apparatus for translating western scientific literature is perhaps on an even larger scale. The scarcity of the world's translation capacity has resulted in research directed to mechanized translation.

The various means of, and the literature on, the propagation of knowledge have evolved from the necessity of solving the tensions between the mass demand for scientific knowledge on the one hand, and, on the other, the growing complexity of acquiring such knowledge and the world-literary character of science. In the socialist countries this work is being carried out within large-scale organizations.

6. *The contrast between the tendency of the scientific information apparatus to expand and the tendency of its relatively decreasing use*

The scientific information organizations and their services can be seen to increase all over the world at a more rapid rate than does their use; hence the tendency of the latter to decrease in comparison to the former. The Washington Conference on Scientific Information has discussed this phenomenon from various angles in connection with the relative decrease in the use of the specialized literature. At the same time the orientation of research workers and scientific information requires even more time. In the field of social sciences there is as yet no such large-scale and concrete registration undertaken as there is in the field of the natural and technical sciences where such problems as the utilization of time devoted to research distributed over various processes and phases of work have been investigated. The degree of utilization of documentation services in this connection and their relation to the other sources and methods of information (oral, direct exchange of experience etc.). The data obtained by the pertinent assessments made so far cannot be used for drawing far-reaching inferences or for generalizations, especially in the field of social sciences requiring a larger amount of written sources than do other disciplines, but they are indicative of the contrast existing between the tremendous pace at which the number of publications and documents grows and the relative decrease in their utilization.

A USA survey, assessing the utilization of time by 1500 chemists carried out by taking samples twice a day picked out at random from ten major cities, found that scientists devote about 50% of their time to gathering and forwarding information, and related activities, and 36% to experiments. Among the sources of information, oral communication dominates, written communication takes up about half as much time, whereas the time devoted to consult publications is again merely half of the latter. An analysis of the utilization of periodicals in 1956 of the London Science Library shows that 4,821 of the 9,120 periodicals were consulted by no one in that year, and the degree of utilization of three quarters of the periodicals was so low that one copy of each would have sufficed for the whole of the country.[15] On the other hand Bernal's classical analysis of the utilization of the scientific literature by research workers testifies to an invariably high rate of consultation. A sample of 208 research workers in eight scientific institutes shows that they consult 5 to 10 periodicals a week and that among the sources of their research papers periodicals amount to 37%, abstracts to 18% and personal talks to 14%. All of those consulted use abstracts and 76% of them regularly read journals.

Or could the growth of documentation be conceived as a 'safety valve' of research, as a means 'to sooth the conscience', to have documentation at hand if necessary without making full use of it? Or is the awareness of its existence, and the mere possibility of its use important? This, would not really account for the large sums assigned to scientific information all over the world.[16]

All this goes to show that the increase of documentation – reflecting the effort to keep pace with the growth of literature – and its decreasing use by research workers cannot be stated unequivocally, it requires further more particularized investigation within the field of each discipline, especially in social sciences.

7. *The paradox that the transmission function of scientific information (between research and primary literature) has the effect of distancing ('estranging, 'alienating') research from the primary sources*

The essential task of scientific information, as that of other kinds of information, is mediation. It transmits the literature to the research workers by its specific methods, means, and services, but it also removes the research workers from the primary sources of communication (books, periodicals, research reports etc.). It tries to replace these by sources of a secondary character, thus contributing to the process of *alienating* the research worker from the literature. The objective cause of this process is the immensity and constant growth of the literature (an annual output of some 250,000 books and some four million papers) which prevents the individual from surveying even a narrower field of his branch of discipline.

The higher the quality of the documentation services, the more they dispence the research worker from the primary sources of communication. The various bibliographies (with annotations of various depth) intended for drawing attention to the best of the literature are compiled with a view to leading the research worker to the original sources, yet the abstracting services of a higher quality, such as thematic documentation surveys compiled according the subjects, have selected, read, compiled the literature of a subject *instead of* the research worker; these 'semi-products' of research i.e. highest-level documentary publications separate him farthest from the primary sources. In fact the process is the same as that which takes place in the selection and abstracting of titles in a directory though that is on a much smaller scale. These abstracts and surveys are not only meant to attract the attention of the research worker to what is the most relevant for him but they sometimes replace reading itself. Yet even the best abstracting service thinks in general terms, that is, it can only select what is of interest for the given field of research. It cannot substitute the research worker who has to solve his own particular problems. And since not even the use of electronic devices can ensure specialized information for individual research workers, the problems of quality can only be solved by working up a large quantity of publications. In other words, documentation takes charge *collectively* of the task of surveying the literature which the *individual research* worker is unable to achieve. In doing so it enhances the possibility of finding the required publications though, at the same time, it plants itself to a certain degree between the research worker and the literature.

Let us mention another paradox under this same heading. The libraries, provided with their available space, try to switch over to what is referred to as the open-shelf system to cut out the mediating function of the catalogue and to make the books and periodicals in classified arrangement directly accessible to the readers. But at the same time attempts are made to introduce punched-card systems and the use of electronic machines which dispense the researchers even of the possibility of looking through the titles or abstracts since the selection is made by the machine. These methods provide a much quicker access to the literature required in various combinations and in great detail, but it remains to be seen whether or not this is indeed the best solution in the field of social sciences for other than very concrete and well-definable themes. If it is, it requires a complete change in the methods of approach to work in these sciences, but it still strengthens the process of 'distancing' ('alienating') the literature.

8. *The contrast between the lack of a general theoretical systematization of sciences and the demand for a documentary classification*

The establishment of a universal and theoretical classification of sciences is still in a stage of research. In the absence of such a classification, a documentary classification must be used to provide research with the necessary information. It is a fact even if the classification of sciences must yield the common basis, because the two systems of classification cannot be identical on account of their purposes and methods. The classification of sciences is of a theoretical character while that of documentation has a practical purpose. In theory the difference between the two classifications cannot be as great as it is in practice. Beside the classification of sciences and the analysis of the 'traditional' classification for documentary purposes, research into 'machine languages' and the experimentation of 'thesauri' is coming more and more to the fore as one of the basic conditions of automating documentation.

9. *The contrast between the interdependency of research and scientific information, and their institutional separation (organization factor)*

What seems to be expedient in principle is that documentation should be located where research is carried on. i.e. the former should be subjected to the latter. This is not the case in practice.

The organizational picture of scientific information is extremely varied throughout the world, it is either organized within libraries or as independent institutions or else attached to some scientific institution. These are the three main organizational types. Their possible variations will now be disregarded: whether the library is part of the documentation centre or the other way round and whether they are co-ordinated and the different ways in which a documentation centre may be attached in different ways to the financing body, which might be a state authority, an institution of higher education, a research institute etc. What we are investigating are not the organizational forms of documentation but the fact that two of the principal types of organization are separated from the workshop of research. We may also disregard the concomitant financial, administrative-economic and other considerations since they do not affect the heart of the question, whether we require a *centralized* documentation, following and registering the scientific development of the world and the pertinent literature, not focussed on the area of coverage of a given institute, or a *decentralized* documentation, focussed on the research demands of one single institution. The answer is that both are necessary, and this is how the separation between the organization of research centres and information centres can be solved in both theory and practice.

By centralized or decentralized documentation we understand whether or not is should be attached to research centres of each branch of science. In this case it is excluded that the documentation of all branches of science should come within the framework of one given organization. The research centres (all kinds of creative workshops of science), are bound by a given programme, by short or long-term projects, consequently, the work of a centre is determined by the short-term project in hand, whence also that of the co-ordinated information unit. But this programme does not necessarily coincide with international scientific development, with the appearance of new disciplines, with the evolution of new research trends, but evolves under the impact of certain considerations (the solution of certain concrete problems or contribution to them, participation in some major complex research project etc.) in wich the above-mentioned factors play only a partial role. With the period for the completion of the research project coming to its end, the research centre will assume new tasks, to study new themes with which the documentation unit must also fall in line. But a new research theme requires earlier literature and formulates its definition. If the documentation, since it has worked according to another programme, is not adequately prepared for this change it is probable that is has not processed the literature necessary for the new task. A subsequent documentation is always a complicated and expensive task. To give an example: the principal theme of an institute is to study the productivity of labour, then by a change in the area of research the theme is transferred to an other institute, and a new subject, i.e. power economy, becomes the principal theme. This change will obviously have an impact on the acquisition and registration of the pertinent literature, too.

Viewed from another angle: research develops at *unequal rates* also in the international field. The application of certain research themes or methods, the setting of new aims necessarily involves certain discrepancies among the various countries. Among other things the co-ordination of these is aimed at by an international division of labour.

Cybernetics or the application of mathematical methods in social sciences has developed a considerable specialized literature, when such research in Hungary, for instance, did not even start and had no organizational forms. Hence there was no 'research centre' to undertake the acquisition and registration of the relevant literature. Before the beginning of research in Hungary the Academy Library started to collect literature on cybernetics, and when research started, the most important manuals and bibliographies, at least, were available. A scientific information centre should register the literature promising to be of interest and use on an international scale independently of the topics studies in the various institutes. Scientific information centres 'keeping an eye' on the scientific developments all over the world, and being relatively independent of current researches is, potentially, a significant factor.

And vice versa, a documentation unit not attached to a research centre is unable to serve the demands of a given institute as specifically as can one which is attached to it. This problem could perhaps be solved if the research centres well aware of the trends in scientific development were to co-operate with the information centres (in the form of scientific advices etc.), and these would operate independently of the themes then under study, whereas the documentation services of the research institutes would contribute to the work of the financing institute. The desirable extent of co-operation and specialization can only be determined concretely and separately in each case, with due regard to the characteristics of the individual disciplines.

10. *The discrepancy between the demand for information experts and its shortage owing to the greater attraction of research (this includes the dilemma of the documentalist: subjective factors)*

Up-to-date research needs specialized documentation, which requires from the documentalist proficiency in the given field of science and, at a certain level, knowledge comparable with that of the research worker. At the same time research work constantly acts as an *absorbing force* attracting the specialists from documentation into the field of research. The higher material and social esteem of research work, the possibility of creative work and of personal satisfaction of scientific ambitions have caused a dilemma for the librarians and, more recently, for the documentalists.

But if all those suitable for scientific work were to engage in higher education or in research, who would be left to satisfy the justified demands made by libraries and for scientific information in general? A librarian's or documentalist's qualification alone is not sufficient. Nor would it, naturally, be convenient if, for one reason or other, a library or documentation unit were to become a shelter of unsuccessful scientists instead of being scientific workshops attracting talented people and ensuring them appropriate prospects. The resolving of this discrepancy will probably take much time, since it is connected with practical problems, such as the granting of more research possibilities for information experts, as well as with problems theoretically not properly elucidated, such as the place of scientific information itself within science.

11. *The discrepancy between the increase in scientific information and the comparative backwardness of its theory*

Only a very small part of the huge number of publications on librarianship and documentation has a theoretical character. But this is a more or less natural proportion, since so many practical problems have arisen over the past twenty years in the course of scientific information work (internal organization, finances, publications, co-ordination etc.) that they have absorbed almost all the attention. But also contributing to this scarcity has been, in all probability, a lack of understanding of the theoretical problems of scientific information – to be more exact, the prevalence of certain opinions, according to which such theoretical problems do not exist, this being an activity that consists purely of practical work. This has then led to psychologically understandable reactions among librarians and documentalists who, when whishing to justify the scientific character of their work, of their profession, started to ascribe such a character to purely practical activities, thereby reinforcing misunderstanding. In addition to this, the situation was further complicated by the disagreement among librarians, documentalists, and bibliographers concerning the relationship between their closely interconnected fields. These circumstances have resulted in that this branch of knowledge and activity engaging many tenthousands of people, directly using up tremendous material means, influencing economic life and scientific work by sums of thousands of millions has, in fact, no established theory and, in many respects, no consolidated higher education.

By achieving a classification of sciences, and by investigating the process of how science is becoming a direct productive force, this tension will be decreased. Information in social sciences contributes to the solution of these two research tasks of a specifically social-scientific character partly by a high-level documentation of pertinent research, partly by raising and theoretically examining its own specific problems within the terrain of the social sciences.

B. THE NATURE OF THE POINTS OF DIVERGENCE BETWEEN RESEARCH AND SCIENTIFIC INFORMATION

The points of divergence analysed in the foregoing are not of the same nature; they are not of identical importance and also differ in our approach to their solution. Some of them seem to be transitional tendencies, others appear to persist, and others again promise a solution within a foreseeable time.

In spite of their differences, it does not seem to be reasonable to assign them to different categories or groups or to separate them, for this would prevent us from recognizing their interrelations and evolving a comprehensive approach to these series of divergences. It is also evident that the solution of certain of these, i.e. the rate and mode of their solution, is not independent of the social conditions in which they arose, and, though generally valid for the social sciences (and even for research and scientific information), have not exactly the same significance in every discipline.

Considering the above, the characterization of the points of divergence can be summed up and completed as follows:

The contradictions related to research demand (1), to the immensity of the body of our knowledge (2), to the exponential increase in the number of scientific works (3), to the process of differentiation and integration of the literature (4), to its 'estranging' (7) could be considered as having a character *sui generis* inasmuch as their effect, though expected to decrease with time, will survive in the form of tendencies, and the same applies to the points of divergence connected with the need for information experts (10). The machine-storing and retrieval of data is expected to considerably curb the effect of the divergence related to the accumulation of knowledge (2). The paradox connected with the exponential 'technical depreciation' of knowledge (3) will have a smaller impact, if any, on the historic and theoretical research, its validity applying mainly to applied and production research. And as far as the need for information experts goes (10), the discrepancy in this respect will always persist. The attractive force of creative work will invariably prevail since all over the world research workers are in increasing demand. The same also applies to information experts. The effect of this dicrepancy could partially be reduced by an increased social and material recognition and stimulation of scientific information as a profession, as scientific work. Such factors as the development of professional training (higher education in two subjects – a discipline and scientific information), the mechanization of documentation and also organizational changes connected with original (primary) publications (cf. Bernal's and Dubinin's suggestions, chapter I) could contribute in diminishing this discrepancy.

The preconditions for a gradual solution of the tension related to the social demand for the increasingly wide application of scientific knowledge (5), are realized by the broadening of the educational system and of adult education (the popularization of science, wide network of libraries, publication policies etc.). The divergences connected with the growth of the information apparatus (6), with the lack of science classification (8), with the organizational factor (9), with the subjective factor (10) and the relative backwardness of the theory of scientific information (11) are likely to be solved, for the most part, in the long range.

The higher the standard of scientific information becomes (mechanization, classification, summary, synthetic surveys etc.), the larger the circle of people who are expected to make use of it (6). The elucidation of the relationship between the classification of sciences and that of documentation (8) and a wide-scale elaboration of the theory of scientific information (11) have been the set aim of a variety of research and here the time factor is completely unknown. World-wide experiments are being conducted to establish the most rational organizational relationship between research centres and information centres (9), and very different reasonable solutions can be conceived as a function of local conditions, of disciplines, and of the concrete demands for science organization.

The points of divergence existing between research and scientific information are part of the problems of science organization, they are of a complex character and, as such, are connected, in more than one respect, with questions of science theory, sociology and economics, and, last but not least, with the basic problem of the theory of scientific information.

APPENDIX A

A 'computer-based' information system: the ISIS, an Integrated Set of Information Systems *

A. WHAT IS THE ISIS?

ISIS is an Integrated Set of Information Systems developed and implemented in the Central Library and Documentation Branch of the International Labour Office in Geneva. It is composed of three interdependent systems: a bibliographic control system, a serials system and a loan system. A fourth evaluation and control system receives information from the other three to permit subsequent analysis. Those modules which are already operational fit into each other like building blocks and the others continue to function manually during the process of systems analysis and programming. The entire set of systems is designed in such a way that it can at any moment be adapted to fit *into a larger system* which might comprise non-documentary as well as documentary elements. By the same token it could be integrated into an information network whose parameters might extend beyond those of the ILO.

* An abreged version of the *ISIS – A short guide for librarians and documentalists,* by G.K. Thompson, W.D. Schieber, D.D. Barrett and G.W. Thomas, Geneva, November 1970, International Labour Office, 16 p., LD/NOTES/50.
The author expresses his gratitude to the Chief of the Central Library and Documentation Branch of the ILO, G.K. Thompson, for his kind permission on utilizing the 'Guide'. The italics and the sub-titles are mine (R. Gy.).

B. HISTORY AND APPROACH

The original idea for ISIS emerged from a series of studies on automation undertaken in the ILO in 1963-64 on behalf of the US Office of Manpower, Automation and Training. At the beginning most efforts were placed on methods of document analysis and techniques of computerised retrieval of abstracts. Early in 1965 the library and documentation services in the ILO were merged as the result of an administrative reorganisation. Since that date ISIS has been gradually implemented, and it emerges today as an integrated *set* of information systems encompassing practically all of the routine activities of the Central Library and Documentation Branch.

In spite of this *global approach,* it was out of the question to attempt to solve all possible philosophical, theoretical or practical problems that such an approach necessarily implies. It was necessary to proceed *pragmatically,* occasionally, riding off simultaneously in several directions, but always adhering to the strict rule that all of the different operations must ultimately be *integrated* into a coherent whole, attempting to ensure that the same information should not be recorded more than once.

It was found necessary to depart from a certain number of *valid models* which could not provide what it was desired to obtain from a modern computerised system. MEDLARS (Medical Literature Analysis and Retrieval System) and a number of other large-scale systems, for example, only offer a listing of subject headings or descriptors out of context, and the MARC (Machine-Readable Cataloguing) system was essentially a mirror of the Library of Congress cards. In these two examples, the systems were designed by librarians essentially for use by other librarians. It was felt that the computer should permit one to go much farther and enable *a direct dialogue with research workers* and other users to take place. It seems reasonable to assume that within a relatively short span of time, many research workers will have display terminals in their offices connected to a computer or a network. After a brief exposure to work with such devices it becomes evident that the research worker himself will probably wish to 'browse' through the recorded abstracts. When this takes place the role of the librarian and documentalist will be seen in a new light. This new role is no less exciting than the one played presently, because it will be the job of the librarian to make this man-machine dialogue possible.

C. BIBLIOGRAPHIC CONTROL SYSTEM

Each of the component systems of ISIS is made up of a certain number of modules, five in the case of the bibliographic control system:

a) acquisitions module;
b) cataloguing module;
c) document analysis module;
d) modification module;
e) output module.

In fact such modules exist in any manual system but are rarely designated as such. The data capture takes place in the first module, and this *unique* recording is used throughout the others. The cataloguer verifies the accuracy of the data already in the computer and makes whatever modifications seem necessary, the document analyst adds the *abstract* used to create inverted files, indexes and other bibliographic products. These various working data are kept in a processing file until final approval had been given to the information recorded for each document. At this point the recording is integrated into the master file.

Data is first recorded on special forms or worksheets, where a separate space is provided for each data field (author, title, and so on). After correction, the worksheets may be transcribed into punched cards for off-line input into the computer or on-line using a visual display terminal, the latter technique being used at ILO for all data recording at present in ISIS.

Field Definition Tables are used by the computer programs to control the creation and correction of new records. The table consists of one entry for each fixed-length element and each variable-length field required in a particular file. The table also

New	Update

* * * ISIS BIBLIOGRAPHIC WORKSHEET * * *

Control Number (00)		1. Document Date (00)		2. Call Number (00)	
3. Personal Author(s) (10)					
4. Corporate Author(s) (11)					
5. Title (20)					
6. Source (30)					
7. Abstract (40)					
8. Language(s) (05)	ENGL FREN GERM RUSS SPAN ITAL Other:			9. Exp. Life (00)	
10. Language Cross Reference(s) (07)		17. Document Number(s) (37)			
12. SBN (04)		13. Price (08)			

PROCESSING CONTROL	Date	Initials		Date	Initials
1. Received for processing 2. Catalogued 3. Analysed			4. Revised 5. Input		

ILO 646 12.70

indicates which fields may be repeated, and in the case of the on-line version, which are requested by the computer of the person keying in the data, and in what order the fields are to be placed in the file. One such Field Definition Table is required for each file to be processed, and this procedure allows for the creation of a great variety of different data files by means of a single computer program.

In the case of the bibliographic file, the on-line data entry routines function as follows:

The recording of bibliographic data takes the form of a dialogue between the computer and the cataloguer (or typist) in the following manner. The computer first requests the sequential control number of the document. When this is typed in, the computer either displays the bibliographic data, if it is a document already placed on the processing file, or asks if it is a new record. If the cataloguer types in 'YES', the computer then asks for a number of compulsory elements that must be present for each bibliographic record. The computer requests these one at a time:

language(s) of the document;

document date, i.e. the date the document was written (not necessarily the same as the publication date);

call number and copy number;

estimated length of time the document should be recorded in the system (one year, two years, five years, etc.);

personal author;

corporate author;

title;

imprint, series notes, holdings notes.

The cataloguer can also request that certain other elements be recorded, such as the International Standard Book Number, the price, document number, cross references to other language editions, etc.

As each element is recorded, the computer can make a certain number of checks to facilitate the 'correction' of the 'manuscript'. Languages, for example, are identified by the first four letters in English (Engl, Fren, Span, etc.). Any non-recognized language will be rejected by the computer. If a name is typed instead of a year, an error message will appear on the screen and so on.

As each element is typed on the keyboard, the same text appears on the screen. Before transmitting any element to the computer, the cataloguer may re-read what has been written. If any errors are detected, they can be located on the screen with a small marker (cursor) and rectified. Once the cataloguer is satisfied that the text is error-free, the 'entry' button is pressed, and the text of that element is sent to the computer. If errors are detected, the computer responds by displaying an appropriate message on the screen, and these must be corrected before the computer will go on to the next element.

It is likely, however, that a number of documentation centres collaborating with the ILO will not have this on-line capability. Very simple punching instructions have been prepared for those centres wishing to continue to prepare punch cards.

Documents chosen for analysis include monographs, internal reports, journal articles, government publications and international organisations documents. Out of a total of approximately 5,000 received, 1,085 journals are regularly scanned for this purpose. *A short abstract of each document is prepared. Descriptors selected from a well-defined vocabulary* are embedded in the text of the abstracts. These descriptors* are contained in *a multi-lingual thesaurus* on economic and social development which has been developed as a co-operative effort of some twenty institutions which are correspondents of the OECD Development Enquiry Service.

The document analyst is instructed to bring out the *main subjects of a document* in the first 100 or 120 characters of text. This first part of the abstract is separated from the rest by a

* *Macro-Thesaurus on Economic and Social Development*, Paris, OECD, 1972.

blank space, a hyphen and another blank space. Each descriptor is delimited by slashes (e.g.,/rural development/) so that the computer can recognize it easily. These slashes are required for input and modification of recorded abstracts, but can be suppressed in printed indexes.

Sample abstract for a monograph

Case study of /economic policy/ and /monetary policy/ measures to achieve /economic development/ and /financial/ stability in /Bolivia/ illustrating the applicability of such anti-/inflation/ary measures in /Latin America/ – examines /obstacle/s to development, factors which contributed to inflation (including the /role of USA//economic aid/ therein), the role of /public enterprise/, /political problem/s, foreign exchange/, /price/ policy, etc. /Bibliography/ pp. 759 to 770 and /statistic/s.

Sample abstract for a journal article

/USA/. Article on /social research/ into the influence of a person's /social status/ on his /social participation/ in /community development/ and /community relation/s in a New York State /rural/ community – covers participation in /volunteer/ /interest group/s, and concludes that it is difficult to cross the boundaries of /social structure/ set by income and the location of residence.

Sample abstract for a government document

Compilation of reports on /demographic/ problems in /France/, with particular reference to means of encouraging /family/s to increase the birth rate – examines the impact of /fiscal policy/ on natality, /housing/ conditions, /family allowance/s, and discusses as well as the /woman worker/ in /part time/ jobs, /immigrant/s and naturalisation, the /social integration/ and /living condition/s of /migrant worker/s, etc. /Statistic/s.

INTERNATIONAL LABOUR DOCUMENTATION

INTERNATIONAL LABOUR OFFICE, CENTRAL LIBRARY AND DOCUMENTATION BRANCH, CH 1211 GENEVA 22

VOL.23 NO.393 ITEMS - 46640-46863 15 FEBRUARY 1972

SUBJECT INDEX

ACTUARIAL
46663 REPORT TO THE GOVERNMENT OF IRAQ ON THE FIRST ACTUARIAL REVIEW OF THE SOCIAL SECURITY SCHEME ADMINISTERED BY THE WORKERS' PENSION AND SOCIAL SECURITY INSTITUTE.

ADMINISTRATION OF JUSTICE
46819 COMUNICACIONES MEXICANAS AL 8 CONGRESO INTERNACIONAL DE DERECHO COMPARADO (PESCARA, 1970).

ADMINISTRATOR
46713 DIRECTORIO DE FUNCIONARIOS DE INSTITUCIONES AMERICANAS DE SEGURIDAD SOCIAL, ORGANISMOS INTERNACIONALES Y DEPENDENCIAS GUBERNAMENTALES VINCULADAS CON ESTA CAMPO.

ADULT
46859 BERUFSBILDUNG ERWACHSENER; AUFGABEN UND LOESUNGEN.

AFRICA
46685 LES CHEMINS DE FER AFRICAINS ET MALGACHES, 9E ED.
46811 REPORT OF THE PANEL OF EXPERTS ON ADVANCED INSTITUTES FOR APPLIED SCIENCE AND TECHNOLOGY IN AFRICA, UNIVERSITY OF MANCHESTER, 10-14 AUGUST 1970.

AFRICA, SOUTH OF SAHARA
46677 L'ENSEIGNEMENT EN AFRIQUE TROPICALE.

AGRARIAN REFORM
46798 MEMORIA DEL PRIMER CONGRESO NACIONAL DE REFORMA AGRARIA, ENERO 5-10, 1970.
46825 TENDENCIA DE LOS PRINCIPALES EFECTOS DE UNA REFORMA AGRARIA EN COLOMBIA.
46826 REFORMA AGRARIA.

AGRICULTURAL POLICY
46826 REFORMA AGRARIA.

AGRICULTURAL PRODUCT
46695 RECENT DEVELOPMENTS IN THE EXPORT TRADE OF PRIMARY PRODUCING COUNTRIES IN AGRICULTURAL COMMODITIES.

AGRICULTURE
46664 NUTRICION Y AGRICULTURA.
46720 THE ECONOMICS OF WORK SHARING IN PEASANT AGRICULTURE.
46731 JORDBRUKSTELJINGA, 20. JUNI 1969. CENSUS OF AGRICULTURE, 20 JUNE 1969. V. 1
46752 AGRICULTURAL DEVELOPMENT IN THAILAND.

AMERICA
46713 DIRECTORIO DE FUNCIONARIOS DE INSTITUCIONES AMERICANAS DE SEGURIDAD SOCIAL, ORGANISMOS INTERNACIONALES Y DEPENDENCIAS GUBERNAMENTALES VINCULADAS CON ESTA CAMPO.

ANDEAN REGION
46699 EMPLEO, DESEMPLEO Y PRESTACIONES POR DESEMPLEO.

ANGOLA
46829 COLOQUIO SOBRE A ECONOMIA DA REGIAO BIENA.

APPRENTICE
46682 LA FORMATION PROFESSIONNELLE DANS L'INDUSTRIE DES MACHINES.

APPRENTICESHIP
46659 SYMPOSIUM ON PART-TIME TECHNICAL AND VOCATIONAL EDUCATION, NOORDWIJK, NETHERLANDS, 29 MARCH-3 APRIL 1971; REPORT.

CHILE
46664 NUTRICION Y AGRICULTURA.
46826 REFORMA AGRARIA.
46831 PLAN DE DESARROLLO DE LA 4 REGION VALPARAISO-ACONCAGUA, 1971-76.

CHINA (R)
46678 THE EFFECTS OF STABLE POPULATION ON ECONOMIC DEVELOPMENT IN TAIWAN.
46733 ESTABLISHMENT SURVEY ON EMPLOYMENT, HOURS, EARNINGS AND LABOR TURNOVER IN SECONDARY INDUSTRIES IN TAIWAN AREA, REPUBLIC OF CHINA (MANPOWER REQUIREMENTS SURVEY), JULY 1971.
46762 DEVELOPMENT OF THE CO-OPERATIVE MOVEMENT IN THE REPUBLIC OF CHINA.

CHINESE
46726 ETHNIC AND SOCIO-ECONOMIC PATTERNS IN MALAYSIA.

CIVIL SERVANT
46713 DIRECTORIO DE FUNCIONARIOS DE INSTITUCIONES AMERICANAS DE SEGURIDAD SOCIAL, ORGANISMOS INTERNACIONALES Y DEPENDENCIAS GUBERNAMENTALES VINCULADAS CON ESTA CAMPO.
46714 ESTUDIO DE PRESUPUESTOS FAMILIARES PARA EL SECTOR DE OBREROS Y EMPLEADOS DE LA CIUDAD DE ROSARIO.
46801 ESTATUDO DEL PERSONAL CIVIL DE LA ADMINISTRACION PUBLICA NACIONAL.
46828 COLLECTIVE BARGAINING IN THE PUBLIC SECTOR; AN INTERIM REPORT, FEBRUARY 13, 1969.

COAL MINER
46669 MINING REDUNDANCY; A CASE STUDY OF THE WORKING OF THE REDUNDANCY PAYMENTS ACT IN THE DURHAM COALFIELD.

COAL MINING
46669 MINING REDUNDANCY; A CASE STUDY OF THE WORKING OF THE REDUNDANCY PAYMENTS ACT IN THE DURHAM COALFIELD.

Abstracts are recorded on the terminal in the same way as the descriptive cataloguing data. It remains to be seen whether the document analysts should record the data themselves directly on the terminal. The computer can verify the validity of descriptors, and could be programmed to detect spelling errors in commonly recurring words. Error checks greatly facilitate the work of the editor of the abstracts, as a fastidious part of his checking work is eliminated.

D. MACHINE PROCESSING OF ABSTRACTS

For more than seven years the principal effort of the ILO library and documentary service has been focused on the improvement of techniques of processing of *machine-readable abstracts.* Each element of a bibliographic reference is tagged so as to permit the generation of a file similar in structure to MARC-II. Some 50,000 such references have been recorded in two versions, one on an IBM 2316 magnetic disc and the second on a security 'back-up' magnetic tape. The disc contains the master file of references and abstracts together with an inverted file used for index generation and in document retrieval. The ILO has developed a special data

Sample card (See opposite page under Agriculture and Chile)

```
46664                        1970                        70A2484
BAYTELMAN D
WISCONSIN. UNIVERSITY. LAND TENURE CENTER
NUTRICION Y AGRICULTURA.
MADISON, 1970. 28 L. TABLE. (ITS: LTC NO. 69-S.)
   ESSAY ON THE /ECONOMIC IMPLICATION/S OF /AGRICULTURE/ IN
   /CHILE/, WITH PARTICULAR REFERENCE TO THE
   INTERRELATIONSHIPS OF /INCOME DISTRIBUTION/,
   UNDER-/NUTRITION/ AND THE /LAND TENURE/ SYSTEM.
ENGL
```

compression technique so that such a large number of abstracts can be accomodated on a single disc.*

Two magnetic discs are used: the master file of 50,000 references with the inverted file referring to it, and a processing file containing information about all documents in the pipeline. This processing file contains the complete status of each item from the time it is ordered until the moment it is transferred into the master file.

Modifications to the 50,000 existing records are also made in the processing file. A record to be corrected is copied from the master file and placed on the processing file where it can be manipulated with various text editing programmes which allow new text to be inserted, or deleted and errors to be corrected. Once modified in this manner the corrected records return to the master file in batches where they replace the earlier version. It is also possible to make large-scale modifications, such as descriptor changes (e.g., Congo (Kinshasa)→ Zaire) directly on the master file.

At the present time new records are transferred from the processing file to the main file once a month. During the transfer operation an alphabetical subject index is printed and the master file updated. This index is reproduced together with the relevant abstracts in a semi-monthly current awareness bulletin entitled, *International Labour Documentation* which appears in paper and card form. The former is widely distributed to research workers within the ILO and to outside recipients. The card edition (3"x5") is used to make up two card catalogues in Geneva. Author cards are supplied to the National Library in Berne, and individual staff members wishing to keep their own manual card files may also request this card edition.

The cards are filed in the catalogues by author, title, subject, country, conference, report number and in the shelf list. As it is virtually impossible to file in the subject catalogue by all of the possible permutations of descriptors, a limited number of key descriptors used as subject headings are underlined; and only these are filed manually.

* Schieber, W, and G. Thomas, 'An algorithm for compaction of alphanumeric data', *Journal of Library Automation*, December 1971.

All bulletins issued from 1965 to 1971 have been replaced by cumulative versions published by G.K. Hall and Co. of Boston.

Copies of those cumulations have been placed in each of the regional offices of the ILO. Decentralized searching is made possible in this manner. Nevertheless in certain circumstances, it may be useful to carry out a search on the computer which results in a printed bibliography. To date some 5,000 such bibliographies have been prepared by the ILO documentation service. Once again, the technique employed is that of a dialogue with the computer via visual display terminals.

E. AN EXAMPLE TO PREPARE A BIBLIOGRAPHY

Let us suppose that we wish to prepare a bibliography of text-books in French on the use of mathematics or statistics in management. The question must be first broken down into a boolean expression, such as:

A+ (B/C) + (D/E/F/G/H) + I.

The search can be initiated by first typing the equals sign (=) to indicate the beginning of a search, followed by, for example, TEXTBOOK. After two or three seconds the computer displays the following message on the screen:

P = 1728 6 = 1728.

The P (Postings) indicates the total number of documents recorded in the system in which the descriptor 'Textbook' occurs. The T indicates the total number of answers to the question as it has been thus far formulated. Within a facet containing 'or' descriptors, opened by a parenthesis, the T refers to the total within that facet only; closing of the facet with another parenthesis provokes a 'real' T (*), resulting from a comparison with the total of that particular facet with the previous total.

At any point in the formulation, except within a facet by parentheses, the computer can be asked to display the abstracts of documents which correspond to the T at that stage of the search. In this manner it can be readily seen whether the formula-

```
=  TEXTBOOK
                   P  = 01728          *  T  = 01728
(  MANAGEMENT
                   P  = 02120             T  = 02120
*  MANAGEMENT DEVELOPMENT
                   P  = 00999             T  = 02870
)
                   P  = 02870          *· T  = 00454
(  STATISTICAL METHOD
                   P  = 00713             T  = 00713
*  MATHEMATIC
                   P  = 00403             T  = 01042
*  NETWORK ANALYSIS
                   P  = 00007             T  = 01047
*  PERT
                   P  = 00044             T  = 01071
*  SIMULATION
                   P  = 00176             T  = 01191
)
                   P  = 01191          *  T  = 00071
+  FREN
                   P  = 08109          *  T  = 00012
```

tion is retrieving relevant documents or whether it is necessary to modify the formulation in some way to get it closer to the required subject. A number of other devices can be used to help 'zero in' on required information.

If the total number of 'hits' seems too high, the requestor may insist that certain descriptors must appear in the first part of the abstract, that is to say before the hyphen separating the main theme of the document from the secondary subjects. This is done by simply reformulating that part of the question by inserting a dot (.) before each descriptor thus designated:

```
(MATHEMATICS
* STATISTICAL METHOD
* SIMULATION
* .NETWORK ANALYSIS
* .PERT
)
```

Such a formulation will unquestionably put the accent on textbooks on PERT.

It is also possible to insert *negative* descriptors, and indicate that no abstracts should be retrieved which contain .NETWORK ANALYSIS. In that case, one should be careful not to eliminate indiscriminately all abstracts containing NETWORK ANALYSIS, as that subject may be only a chapter of a book which should have otherwise been retrieved.

It is also possible to search for *documents written* before or *since a certain date,* or for abstracts published since a certain number, to update a previous search. Even more significantly, the computer can be asked to conduct a free text search for any string of characters present in the bibliographic record, without being limited to the tagged descriptors. In the formulation mentioned above, one could ask for only those works published by Dunod. The formulation would be:

TEXT 30 ' DUNOD '. (30 is the tag number for the imprint)

There are a number of inherent traps in free text searches. When answering a request for information on 'mobile training centres' (mobile is not a descriptor), the following formulation was used:

= TRAINING CENTRE
TEXT 40 ' MOBILE '. (40 = Abstract tag)

The computer found all of the relevant documents, but also quite a number of training centres for auto*mobile* mechanics! Since we had not indicated that we wanted the *word* 'mobile', it simply had searched for the string of characters M-O-B-I-L-E. Of course, it is easy to avoid such a trap by inserting a blank space before the word:

= TRAINING CENTRE
TEXT 40 ' MOBILE '.

Authors' names, words in the title, any element present in the record may be used as a search criterion. In this way it was possible to search for articles on French-speaking Africa which had been indexed in the journal Economic Development and Cultural Change.

As the search criteria are completely open-ended, new methods are being discovered constantly for digging out information. The weighting feature (searching for .NETWORK ANALYSIS instead of just NETWORK ANALYSIS) is one of these. The document analyst weighted certain descriptors anyway by placing them in the first part of the abstract, and this device is most useful in retrieval for helping weed out items of lesser 'importance'. Certainly many other new search criteria will be thought of by the documentation staff as time goes by.

A fuller discussion of these techniques, together with a description of the library routines which form an integral part of the system was published in 1971.*

F. SOME CONCLUSIONS

An international 'computer-based' documentation network can be envisaged. When an institution adopts a standardised format for recording its date, it is not only in a position *to exchange* these data with other institutions, but *may use computer programmes (software) prepared elsewhere* – even if these programmes are written in another programming language or for another computer.

The MARC-II format can unquestionably be amplified for use in recording *non*-bibliographic data. File of personnel data, directors of institutions and research projects data, for example, can be processed with the same computer programmes written to deal with bibliographic data. This remarkable *flexibility* leads one to reconsider the parameters of the library and documentation profession, and to contemplate the consequences of the integration of one's own 'ISIS' in a powerful set of information systems, the full implications of which it may be hard to grasp at the present time.

* Schieber, W., *ISIS; a General Description of an Approach to Computerized Bibliographic Control'*, Geneva, ILO, 1971, 115 p.

APPENDIX B

Bibliography

An introduction to the source of specialized literature of science organization

A. BIBLIOGRAPHIES ON SCIENCE ORGANIZATION

To show the rate of growth and variety of the literature of the science organization we quote some dozens of titles. This selective bibliography of bibliographies is far from exhaustive, but illustrates well the increase of the specialized literature and the growing interest for the problems of the 'science of science'.*

1. *Bibliographies*

An attempt at a bibliography of bibliographies in science of science. Lund, Research Policy Program, 1966, 69 p.

Bibliography on research administration, management, organization and use. Houston, Science Inf. Assoc., 1960, 23 p.

Bóna, E. (ed.), *Tudománytani bibliográfia.* [Bibliography on science of sciences] Veszprém, Vegyipari Egyetem, 1968, 424 p.
5,000 bibliographical items; a periodical list with 500 titles; table of contents in English and Russian.

Bush, G.P., *Bibliography on research administration (annotated).* Washington, Washington Univ. Press, 1954, 146 p.

Bush, G.P., and Hattery, L.H., *Scientific research, its administration and organization.* Washington, The American Univ. Press, 1950, 190 p.

Cairl, J., and Gallagher, R., 'Government, science and technology: A bibliograpical essay', *Public Administration Review,* July, 1968, 373-381.

* For a more complete bibliography (which was used partially on establishing this one) see: *Science policy in ECE countries.* Reference lists no. 4, chap. VI. Geneva, UN Library, ST/GENEVA/LIB/SER.B/Ref. 4, 1971. See also Salomon, J.-J., *Science et politique.* Paris, Ed. du Seuil, 1970; Piccard, P.J., 'Science and public policy: more literature for a field', *Public Administration Review* 225-228, September 1966.

Caldwell, L.K. *Science, technology and public policy. A selected and annotated bibliography 1945-1965.* Bloomington, Inst. of Public Administration, Indiana Univ., 1965. Various pages.

Caldwell, L.K. *Science, technology and public policy. A selected and annotated bibliography.* Bloomington, Indiana Univ., 1968-69. Vol. 1: Books, monographs, documents, 492 p.; Vol. 2: Articles, 544 p.

Ekonomika i organizatsiia nauchno-issledovatel'skoi raboty; rekomendatel'nyi ukazatel' literatury [Bibliography on economics and on organization of scientific research] Moskva, Kniga, 1969, 107 p.

Goldschmidt. D. (ed.), *Social science on higher education and universities. Part II: Annotated bibliography.* Comp. by W. Nitsch and W. Weller. The Hague – Paris, Mouton, 1970, XXVII + 802 p.
4,400 bibliographical items subdivided in 7 sections; a periodical list with ca. 1,000 titles.

Kaplan, N. *Science and society.* Chicago, Rand McNally, 1965. Bibliography: 581-595.

NSF *(National Science Foundation), A selective bibliography on research and development and its impact on the economy.* Supt. of Docs, Washington, 1958, 21 p.

NSF, *Bibliography on the economic and social implications of scientific research and developemnt.* Supt. of Docs, Washington, 1959, 52 p.

NSF, *Current project on economic and social implications of scientific research and development.* Washington, 1961, 124 p.

Nelson, R.R. 'The economics of invention: A survey of literature'. *J. Business,* 32, April 1959, 101-127.

OECD Library, *L'écart technologique entre les pays membres de l'OCDE* [The technological gap between OECD members' countries]. Paris, 1968, 150 p. Special annotated bibliography.

OECD, *La prévision technologique* par E. Jantsch. Paris, 1967. Bibliography: 353-419.
One of the most important bibliographies on the technological forecasting; 413 items in 15 chapters.

OECD, *Problèmes de politique scientifique.* Paris, 1968. Bibliography: 193-215.

OECD, Programme d'échange de documents en matière de politique scientifique. *Bibliographie 1966,* 1966, 241 p. with Index 21 p.

Retting, R.A., *Bibliography on science and world affairs.* Washington D.C., U.S. Dept. of State, 1964, 179 p.

Rhenman, E., and Svensson. S., *Research administration. A selected bibliography of recent literature,* 2nd ed. Stockholm, Aktiebolaget Atomenergi, 1961, 57 p. AE-28.

Richta, R., *La civilisation au carrefour.* Paris, Anthropos, 1969. Bibliography: 429-463, and many bibliographical references in the footnotes.

Russo, F., *Elements de bibliographie de l'histoire des sciences et des techniques,* 2me éd. refondue et augmentée. Paris, Hermann, 1969, 214 p.

Salomon, J.J., *Science et politique.* Paris, Ed. du Seuil, 1970. Bibliography: 376-388, and many bibliographical references in the footnotes.

Science of Science Foundation, Library Committee, *The science of science: An introductory reading list.* London, 1968, 31 p.

Silverberg, B., *Läromedel teknisk undervisning* (Technical progress). Stockholm, Skolvärlden, 1969, 111 p.

Smith, L.R., *The concept of scientific choice: A brief review of the literature.* Santa Monica, Rand Corp., Report P. 3156, 1965, 54 p.

Spaey, J., *Le développement par la science.* Paris, Unesco, 1969. Bibliography: 199-204.

Unesco. *Etude bibliographique comparative sur les dépenses consacrées à la recherche scientifique technique de 1957 à 1963.* Paris, Unesco, NS/ROU/24, 1963, 25 p.

Unesco, *Etude bibliographique comparative sur les ressources humaines en science techrologie. Situation actuelle.* Paris, Unesco, NS/ROU/28, 1963, 83 p.

Unesco, *Main trends of research in the social and human sciences. Part I: Social sciences.* Paris/The Hague, Mouton/Unesco, 1970, XLVII + 819 p.
French edition: *Tendances principales de la recherche dans les sciences sociales et humaines. Première partie: Sciences sociales.* Paris – The Hague, Mouton/Unesco, 1970, LII + 987 p.
With very relevant bibliographic material.

Unesco, *Liste des documents de la série Unesco/NS/ROU.* Secteur des sciences: Division de la politique scientifique. Paris, Unesco, NS/ROU/77/Rev. 1, 1969.

UNIDO, *Indistrial programmimg section.* Vienna, UNIDO, ID/WG3/BP3, 1967, 23 p.
Selected bibliographical references.

UNECE (UN Economic Commission for Europe), *Seminar on technological forecasting, Warsaw, 1970. Bibliography on technological forecasting.* Geneva, 1970, 9 p.
Its: (Documents) SC. TECH/B-1/Working paper 4, and Add. 1. Roneo.

UNECE, *Science policy in ECE countries.* Geneva, 1971.
Reference lists no. 4 ST/GENEVA/LIB/SER.B/Ref. 4 (SC.TECH. 1971/7).

US Library of Congress, Legislative reference service, *An inventory of congressional concern with research and development. A bibliography.* Washington, D.C., US GPO, 1966, 120 p.

US Library of Congress, *A national program of institutional grants for science and science education.* Washington, D.C., US GPO, 1968, 136 p.

US Library of Congress, *Policy planning for technology transfer.* Washington, D.C., US GPO, 1967.

US Library of Congress, *Science, technology and American diplomacy. A selected annotated bibliography.* Comp. by G.J. Knezo, Washington, D.C., US GPO, 1970, 69 p.

2. *Documentation journals (serials)*

Anotavana bibliografie literatury o organizaci . . . vedeckeho vyzkumu [Annotated bibliography of literature on organization . . . of scientific research]. Praha, 1967 – CSAV (Academy of Sciences).

Bibliographie-Bibliography. Bruxelles, 1962, No. 1. Irreg. Conseil national de la politique scientifique. Bulletin analytique des ouvrages de politique scientifiquc.

Bibliography on social science policy. Langborough (Gr. Br.)., Univ. of Techn. Centre for the Utilization of Social Science Research. Irreg.

BIRPI. Liste bibliographiques mensuelles de la Bibliothèque des BIRPI. Geneva, Monthly (WIPO-World Intellectuel Property Organization).

Documentation bulletin on future research 2000. Groningen, 1970. 10 times a year.

FID. News bulletin. Fédération Internationale de Documentation. The Hague. Monthly.

Informacni zpravodaj. Prehled vybvane literatury . . . organizace vyzkumi. Praha, Ustav planovani vedy CSAV.
Irreg. [Informations on science organization].

Novaia literatura o nauke i nauchno-issledovatel'skoi rabote zu rubezhom. Moskva, 1947. Fundamental'naia biblioteka obshchestvennykh nauk, Akademiia nauk SSSR.
Monthly. [International bibliography on 'science of science']. *

Novaia sovetskaia literatura o nauke i nauchno-issledovatel'skoi rabote v SSSR. Moskva, 1934-. Fundamental'naia biblioteka obshchestvennykh nauk, Akademiia nauk SSSR.
Monthly. [Bibliography on 'science of science' in URSS].*

Phase zéro. Paris, 1971-. Monthly.

Przeglad informacji o naukoznawstwie. Warszawa, 1966-. Polska akademia nauk, Osrodek dokumentacji i informacji naukowej. Quarterly. [Documentation in 'science of science'].

Science information notes. Reporting national and international developments in scientific and technical information dissemination. Washington, National Science Foundation.
Monthly.

Science policy bulletin. Columbus, Ohio, 1967-. Battelle Memorial Institute. Bi-monthly.

Science policy information. Paris, 1967-1969. OECD, Directorate for scientific affairs.
Irreg.

* Barykina, O.A.: *Bibliotechno-bibliograficheskaya rabota po voprosam nauki i ee organizatsii* [Bibliographic work on science organization]. Trudy BAN i FBON, Vol. VI. 1962, 5-47.

Tudományszervezési tájékoztató. Budapest, 1961-. Magyar Tudományos Akadémia Könyvtára.
Bi-monthly.[Documentation on science organization].* With English and Russian summaries.

B. A SELECTED LIST OF PERIODICALS ON SCIENCE ORGANIZATION

An international survey of the literature on science theory, science policy and research organization shows that research organization develops its own periodical literature in accordance with 'research industry'. Some of them are entirely devoted to science organization problems.

1. *General reviews on science policy*

Bulletin d'informations scientifiques et techniques du Commissariat à l'Energie Atomique, Paris. 1957-. Monthly.
Cahiers. Sér. AI. Politique d'orientation de la recherche scientifique et technique. Institut de Science Economique Appliquée, Paris. 1962-. Irreg.
Cahiers. Sér. T. Information, recherche, innovation. Institut de Science Economique Appliquée, Paris. 1959-. Irreg.
Industrial research and development news. United Nations Industrial Development Organization, Vienna. 1966-. Quarterly.
Information scientifique. Ministère des Affaires Etrangères, Paris. Irreg.
Informazione scientifica. Notiziario di scienza e di technica per la stampa. Rome. 1955-. Semi-monthly.
Itogi nauki. VINITI, Moscow. Irreg.
(The results of science, a collection of the Institute of Scientific Information).
Magyar tudomány. Hungarian Academy of Sciences, Budapest. 1956-. Monthly.
Minerva. A review of science, learning and policy. Committee on Science and Freedom, Londen. 1962-. Quarterly.
Nauka polska. Polish Academy of Sciences, Warsaw. 1956-. Quarterly.
News report of the National Research Council. National Academy of Sciences, Washington. 1951-. Ten issues per year.
New technology. News of production, research and development from the Ministry of Technology, London. 1967-. Monthly.
Predpoklady rozvoje vedy a techniky. Ustav pro technicke a ekonomicke informace, Prague. 1963-. Ten issues per year.

* Rózsa, Gy.: *Literature on information on science organization in Hungary*. Zagadnienia Naukoznawstwa, 1968. No. 2. 127-130.

Le progrès scientifique. Délégation Générale à la Recherche Scientifique et Technique, Paris. 1965-. Monthly.

R + D. Research and development for industry. London. 1961-. Irreg.

Research management. Industrial Research Institute, New York. 1958-. Bi-monthly.

Reviews of national science policy. OECD, Paris. 1965-. Irreg.

La ricerca scientifica. Rivista del Consiglio Nazionale delle Richerche, Rome. 1930-. Monthly.
Sub-series: (1) Notiziario; (2) Quaderni; (3) Supplemento.

Scandinavian research information notes. Scandinavian Council for Applied Research, Stockholm. 1966-. Semi-annually.

Science and technology. An interdisciplinary monthly magazine for professional scientists and engineers, New York. 1962-. Monthly.

Science policy and organization bulletin. Department of Education and Science, London. 1969-. Bi-monthly.

Science policy news. Science of Science Foundation, London. 1969-. Bi-monthly.

Science policy studies and documents. Unesco, Paris. 1965-. Irreg.

Scientia. Rivista internazionale di sintesi scientifica, Milan, 1970-. Monthly.

Scientific research. Philadelphia. 1966-. Monthly.

Scientific world. World Federation of Scientific Workers, London. 1957-. Quarterly.
Multilingual editions (English, French, German, etc.).

Spektrum. Mitteilungsblatt für die Mitarbeiter. Akademie der Wissenschaften, Berlin (GDR), 1955-. Irreg.

Spisanie na Bulgarskata akademiia na naukite. Bulgarian Academy of Sciences, Sofia. 1953-. Three issues per year.

Technological forecasting. New York. 1969-. Quarterly.

TVF. Technisk-vetenskaping forskning. Stockholm. 1958-. Eight issues per year.
(Scientific-technological research).

Vestnik akademii nauk SSSR. Academy of Sciences of URSS, Moscow. 1931-. Monthly.

Vestnik Ceskoslovenská akademie ved. Academy of Sciences of Czechoslovakia, Prague. 1892-. Bi-monthly.

Wirtschaft und Wissenschaft. Nachrichten, Beiträge, Informationen. Stifterverband für die deutsche Wissenschaft, Essen – Bredeney. 1953-. Bi-monthly.
(Economy and science).

Zagadnienia naukoznawstwa. Studia i materialy. Polska Akademia Nauk., Komitet Naukoznawstwa (Polish Academy of Sciences), Warsaw. 1965-. Quarterly.
(Studies on science of science).

2. *Natural and social science journals with special regards to science policy*

Advancement of science, The. British Association for the Advancement, London. 1939-. Quarterly.

American scientist. Published in the interest of scientific research. Society of the Sigma XI, Scientific Research Society of America, New Haven (Conn.). 1913-. Quarterly.

Archives internationales d'histoire des sciences. Paris. 1919-. Quarterly.

Atomes. Paris. 1946-. Monthly. (See *Recherche, La*)

British journal for philosophy of science. London. Monthly.

Bulletin of the atomic scientists. A journal of science and public affairs. Education Foundation for Nuclear Science, Chicago. 1945-. Monthly.

Chemical and engineering news. American Chemical Society, Washington, D.C., 1923-. Weekly.

Daedalus. Journal of the American Academy of Arts and Sciences.

Impact of science on society. Unesco. Paris. 1950-. Quarterly.

International associations/Associations internationales. Union of International Associations, Brussels. 1949-. Monthly.

International organization. World Peace Foundation, Boston. 1947-. Quarterly.

International social science journal. Unesco, Paris. 1949-. Quarterly.

Nature. A weekly journal of science. London. 1869-. Weekly.

New scientist. London. 1956-. Weekly.

Note d'information. Centre Européen de Coordination de Recherche et de Documentation en Sciences Sociales, Vienna. 1965-. Bi-monthly.

Organon. Institut d'Histoire de la Science et de la Technique, Académie polonaise des Sciences, Warsaw. 1964-. Annual.
Issued in co-operation with the Division d'Histoire des Sciences of the Union Internationale d'Histoire et de Philosophie des Sciences.

Peace and the sciences, International Institute for Peace, Vienna. 1964-. Quarterly.

Philosophy of science. Philosophy of Science Association, Baltimore (Md.). 1934-. Quarterly.

Recherche, La. Paris. Monthly (Formerly *Atomes,* 1946-.)

Reports and papers in the social sciences. Unesco, Paris. 1955-. Irreg.

Review. International Council of Scientific Union (ICSU), Amsterdam. Quarterly.

Revue internationale des sciences sociales. Unesco, Paris. 1949-. Quarterly.

Science. American Association for the Advancement of Science, Washington, D.C. 1880-.

Science and society. New York. 1936-. Quarterly.

Science progress. A quarterly review of current developments in science. Oxford. 1906-. Quarterly.

Sciences et l'enseignement des sciences. Revue française des sciences et techniques. Paris. 1959-. Bi-monthly.

Scientific American. New York. 1845-. Montly.

Social science information/Information sur les sciences sociales. International Social Science Council (ISSC), Paris. 1954-. Bi-monthly. Bilingual.

Studium generale. Zeitschrift für die Einheit der Wissenschaften im Zusammenhang ihrer Begriffsbildungen und Forschungsmethoden. Berlin (GFR). 1947-. Monthly.

Synthèses. Brussels. 1946-. Monthly.

Wissenschaft und Weltbild. Zeitschrift für die Grundfragen der Forschung. Vienna. 1948-. Quarterly.

3. *Higher education journals with special regards to science policy*

Deutsche Universitätszeitung. Hochschul-Dienst Informationen aus dem wissenschaftlichen Leben. Bonn. 1945-. Semi-monthly.

Felsöoktatási szemle. Müvelödésügyi Minisztérium, Budapest. 1952-. Monthly.
(Review of higher education).

Higher education and research in the Netherlands. Netherlands Universities Foundation for International Co-operation, The Hague. 1957-. Quarterly.

Das Hochschulwesen. Wissenschaftspolitische Rundschau. Ministerium für Hoch- und Fachschulwesen der DDR, Berlin (GDR). 1953-. Monthly.

Osterreichische Hochschulzeitung. Notring der wissenschaftlichen Verbände Österreichs, Vienna. 1949-. Fortnightly.

Revue de l'enseignement supérieur. Ministère de l'Education Nationale, Paris. 1956-. Quarterly.

Schweizerische Hochschulzeitung/Revue universitaire suisse. Zürich. 1927-. Bi-monthly.

Universities quarterly. London. 1946-. Quarterly.

Vestnik vysshei shkoly. Ministerstvo Vysshego i Srednego Spetsial'nogo Obrazovaniia, Moscow. 1940-. Monthly.
(Review of higher education)

Vysoká skola. Orgán Státního Vyboru pro Vysoké Skoly, Prague. 1953-. Monthly.
(Review of higher education)

Zycie szkoly wyzszej. Ministerstwo Szkolnictwa Wyzszego, Warsaw. 1953-. Monthly.
(Review of higher education)

4. *Annual reports of scientific bodies*

American Council of Learned Societies. Annual report. New York. 1957-.
Centre National de la Recherche Scientifique. Rapport d'activité. Paris. 1958-.
Conseil National de la Politique Scientifique. Rapport annuel. Brussels. 1960-.
Deutsche Akademie der Wissenschaften zu Berlin, Jahrbuch der -. Beitrage zur Wissenschaftsorganisation. Berlin (GDR). 1939-.
Koninklijke Nederlandse Academie van Wetenschappen, Jaarboek der -. Amsterdam.
National Academy of Sciences, Report of the -. Washington, D.C. 1918-.
National Science Foundation (NSF), Washington, D.C.
- Annual report
- Current projects on economic and social implications of science and technology. 1959-.
- Federal funds for research development and other scientific activities. 1953-.
- Research and development in industry. 1957-.
- Reviews of data on science resources. 1964-.
- Scientific manpower. 1956-.
- Scientific research and development of non-profit organizations. 1957-.
- Scientists and engineers in the Federal Government. 1961-.

The Royal Society, The Yearbook of -. London. 1896-.

Notes

NOTES TO 'INTRODUCTION'

1. Of this about one and a half billion in the developed industrial areas (USA, USSR, Europe and Japan), and four and a half billion in the developing countries. *See:* Maurice Guernier, 'Le "livre noir" du Tiers Monde', *Le Monde,* Jan. 5, 1972.
2. René Maheu, *Science et synthèse,* Paris, 1967.
3. This work is based on the volume of the author *Information problems of social science research and of science organization* (published in Hungarian), Budapest, Publishing House of the Hungarian Academy of Sciences, 1965, involving its structural transformation and essential development.

NOTES TO 'THE SOCIO-HISTORICAL EVOLUTION OF SCIENTIFIC INFORMATION PROBLEMS'

1. Marx, *Theories of surplus value,* Part 3. London, Lawrence and Wishart, 1969
2. Engels, *The situation of England.*
3. As formulated by A. Kosygin, *Izvestiya,* 15 June 1961.
4. Zvorykin, A., 'Nauka i proizvodstvo' [Science and production], *Kommunist* 4: 36-45, 1962.
5. Marx and Engels, *Selected works.* London, Lawrence and Wishart, 1968.
6. Some references to the thesis 'science as a productive force' literature:
   Auerhan, I., 'Veda i jako vyrobni sila' [Science as a productive force], *Politicka Ekonomie* 3: 185-198, 1962;
   Roman, V., 'Transformarea stiintei in forta nemijlocitá de productie' [Transformation of science into direct productive force], *Lupta de clasá* 11: 53-62, 1962;
   Klotz, H. and Rum, K., 'Uber die Produktivkraftwissenschaft' [About science as a productive force], *Einheit* 2: 24-31, 1963;

Kosel, G.: 'Produktivkraft-Wissenschaft' [Productive force-science], *Die Wirtschaft.* Berlin, 1957. p. 146;
and a fundamental reference: 'Iz neopuklikovannyh rukopisei K. Marksza' [From K. Marx's unpublished manuscripts], *Bolshevik 11-12,* 1939.

7. Marx, *Theories of surplus value,* Part 1. London, Lawrence and Wishart, 1969, p. 288.
8. Molnár, E., *Dialektikus materializmus és társadalomtudomány* [Dialectical materialism and social sciences]. Budapest, Kossuth, 1962. p. 116; for the relation between natural and social sciences see:
Kendall, M.G., 'Natural law in social sciences', *Journal of the Royal Statistical Society,* Series A: 124, (1): 1-19, 1961;
Machlup, F., 'Are the social sciences really inferior? ', *The Southern Economic Journal* 3: 173-184, 1961.
Both authors contest the 'difference in rank' between the natural and the social sciences.
9. See the proceedings of the 1962 general assembly of the Academy of Sciences SSSR in the *Vestnik Akademii Nauk SSSR.*
10. For the data see: *Current trends in scientific research.* New York-Paris, United Nations-Unesco, 1961. p. 246;
Dubinin, M.M., 'Vashnye voprosy obmena nauchnoi informatsii' [Exchange of scientific information], *Vestnik Akademii Nauk SSSR* 4: 40-43, 1962;
Kent, A., 'Machine literature searching in science', *Journal of the Franklin Institute* 7: 42-50, 1960.
11. Addition to the 1956 Russian translation of *Science in history.* London, 1954, p. 681.
12. Kent, A., 'Resolution of the literature crisis in the decade 1961-1970', *Research Management* 1: 49-58. 1962:
Humphrey, H., 'Unknowing duplication in research, a perennial tragedy', *The Trend in Engineering at the University of Washington* 4: 1-2, 1961.
13. *World list of social science periodicals,* 3rd edition. Paris-The Hague, Mouton, 1967. Periodical additions are made in *Social Science Information.*
14. *Proceedings of the International Conference on Scientific Information, Washington, D.C., November 16-19, 1958.* Washington, D.C., National Academy of Sciences, National Research Council, 1959, Vol.1, p. 812; Vol. 2, p. 813-1635.
15. Ames, A., 'Research, invention, development and innovation', *American Economic Review* 51: 370-380, 1961.
16. Puranik, K.D., 'Field of knowledge and its repercussion on classification', *Abgila* 2 (1): 19-23, 1951.
17. Yefreinov, Y.V., Kosarev, V.B. and Ustinov, V.A., 'Vychislitel'naya tekhnika v istoriko-filologicheskihk issledovaniyakh. Analiz drevnikh

rukopisei maya s pomoshch'yu elektronnoi mashiny' [Computation techniques in historical-philological research. An analysis of Maya manuscripts with the help of an electronic machine], *Vestnik Akademii Nauk SSSR* 1: 80-83, 1962.

18. 'An electronic analogue for an economic system' *Electrical Engineering* 4, 1952;
'Sotsiologiya i kibernetika' [Sociology and cybernetics], *Voprosy Filosofii* 5, 1958;
'A kibernatika Marxot igazolja' [Cybernetics justify Marx], *Népszabadság* 4, Oct. 1958.
19. Professor Bernal published his proposals first in the form of theses: *Provisional scheme for central distribution of scientific publication.* The Royal Society Scientific Information Conference, 1948, June 21-July 2. Report and papers submitted. London, The Royal Society, 253-258; see also: 'Vers une révolution technique dans la documentation? ', *Revue de la Documentation* 4: 137-140, 1957.
20. Berkov, P.N., *Bibliograficheskaya yevristika. K teorii i metodike bibliograficheskikh razyskanii* [Bibliographic heuristics. Contribution to the theory and methodology of bibliographic retrieval], Moscow, Vsesoyuznaya Knizhnaya Palata, 1960, p. 175.
21. Pólya, G., 'Les méthodes dans les sciences modernes', *Science et Industrie.* Paris, 1961, p. 279-286.
22. *Proceedings of the International Conference on Scientific Information, Washington. Op. Cit.*, Vol. 2, p. 1523-1533;
King, A., 'Some reflections on the International Conference on Scientific Information, Washington', *Revue de la Documentation* 1: 1-5, 1959.
23. 'Federal funds for science'. Washington, National Science Foundation. *NSF:* 63-11.
24. *Proceedings of the International Conference on Scientific Information, Washington. Op.cit.*, Vol.1, p. 498-510.
25. Poluskin, V.A.: 'Desyat let raboty VINITI' [Ten years of VINITI], *Vestnik Akademii Nauk SSSR* 3: 127-128, 1963.
26. *Proceedings of the International Conference on Scientific Information, Washington. Op.cit.*, Vol. 2, p. 1523-1533;
see also Gresser, K., Paschen, H. and Schwachow, *Die Kosten der wissenschaftlichen und technischen Information* [The cost of the scientific information]. München-Pullach, Verlag Dokumentation, 1970. p. 136.
27. Engels, *An outline for a critique of national economy.*
28. *The Royal Society Scientific Information Conference. Op.cit.;*
Bourgeois, P., 'L'avenir du périodique scientifique', *Libri* 7: 71-75, 1956;
Csapodi, Cs., 'L'avenir des périodiques scientifiques', *Revue de la Documentation* 3: 79-81, 1958.

29. Dubynin, M.M., *Op.cit.* cf. n.10.
30. *American Economic Review* 51, (2), 1961. Papers and proceedings . . Report of the managing director for the year 1961.
31. Such investigations can look back upon a certain history in the literature of the natural sciences and engineering, but have little to go on in the social sciences themselves. The first significant study in this respect was published by P.L. Gross and E.M. Gross on their methods of 'measuring the value' of chemical periodicals: *Science* 66, 1927.
    The method here described was later extended to other fields. Let us quote more recent articles:
    Raising, L.M., 'Mathematical evaluation of the scientific serial. Improved bibliographic method offers new objectivety in selecting and abstracting the research journal', *Science* 1960, May 13, p. 1417-1419;
    Baker, D.B., 'Growth of chemical literature. Past, present and future', *Chemical Engineering News,* July 17, 1961, p. 78-81;
    Rózsa, Gy., *Les sciences sociales hongroises dans les publications de l'Unesco* [Publications Bibl. Acad.Sci.Hung. 20]. Budapest;
    Price, D.J., 'Networks of scientific papers', *Science* 149: 510-515, 1965.
32. ISO/R 214: *The recommandation of the International Standards Institute on author's abstracts.* Paris, Unesco, NS/177: *Guide pour la rédaction de résumés d'auteurs.* Paris, Unesco, NS/51.
33. A few of the relevant general works are indicated as illustration:
    Index bibliographicus. Vol. 2: *Social Sciences.* The Hague, FID, 1964, 4th ed.;
    Berkov, P.N., *Bibliograficheskaya yevristika* [Bibliographical heuristics]. Moscow, 1960;
    Malclès, L.M., *Cours de bibliographie.* Vols. 1-3. Genève/Lille, 1954;
    Kirpicheva, I.K., *Bibliografiya v pomoshch'neuchnoi rebote* [Bibliography at the service of scientific work]. Leningrad, 1958.
    These can be used as introductory bibliographical guides.
    In the field of social sciences the following selection will illustrate the variety and diversity of the themes:
    Research method in social relations. New York, 1959;
    Social science research and libraries. Bombay, 1960;
    Frey, F.W., *Survey research on comparative social change – A bibliography.* Cambridge, Mass., 1969;
    White, C.M., *Sources of information in the social sciences.* New York, 1964;
    Lewis, P.R., *The literature of the social sciences. An introductory survey and guide.* London, 1960;
    Thompson *Gateway to the social sciences.* New York, 1959, rev. ed.;
    Mason, I.B., *Research resources – Annotated guide to the social sciences.* Santa Barbara, Calif., 1968.
    Beside such introductory, methodological and bibliographical works on

social sciences as those quoted above, we can cite a few works in the field of economic sciences:
Coman, E.T., *Sources of business information.* New York, 1949;
Manley, M., *Business information – How to find and use it.* New York, 1955.

34. For the first contemporary survey, see: *Preliminary analysis of pilot questionnaire on the use of scientific literature.* The Royal Society Scientific Information Conference. *Op. cit.:* 101-102, 589-637;
see also the papers submitted to the FID Congress in Buenos Aires, September 21-24, 1970: *Users of documentation – Abstracts of papers.* The Hague, FID 465.
35. For example, on the occasion of the Vth International Congress of Sociology at Evian (see ICSSD Secr. 63/11), and Harvey, I.M., *Information methods of research workers in the social sciences.* London, The Library Association, 1961. 28 p.
36. With a special emphasis on Kedrov, B.M., *Klassifikatsya nauk* [Classification of sciences], Vol. I. Moscow, 1961.
For a summary of social science classification and on mechanization experiments in many respect see:
De Grolier, E., *Etude sur les catégories générales applicables aux classifications et codifications documentaires.* Unesco, 1962;
Foskett, D.J., *Classification and indexing in the social sciences.* London, 1963, 190 p.;
Atherton, P. (ed.), *Classification research. Proceedings of the 2nd International Study Conference (Elsinore, September 1964).* Copenhagen, Munksgaard, 1965, 563 p.
37. Lickert, H.: 'Dokumentation der Wirtschaftswissenschaften in Sicht? ' [Economic documentation in sight?], *Wirtschaftswissenschaft* 7: 1072-1075, 1961.
38. *Bibliotechno-bibliograficheskaya klassifikatsiya. Tablitsy dlya nauchnykh bibliotek* [Library and bibliographical classification]. Moscow, 1960;
For a summary of relevant research conducted in the Soviet Union, see: Shamurin, E.I., *Geschichte der bibliothekarisch-bibliographischen Klassifikation* [History of library classification]. Bd. 1-2. (Translated from the Russian). Leipzig, Bibliographisches Institut, 1964/65.
39. The history of science offers more than one example where applied or development research led also to some significant theoretical discovery, since these three levels of research are in constant interaction. It is theoretically possible that research into a universal documentary classification system can contribute to the successful solution of the research problem of classifying and demarcating in the sciences.
40. In connection with UDC criticism and revision we quote
Fill, K., 'Kritische Gedanken zur Revision der Dezimalklassifikation' [Criticism on the revision of the UDC], *Revue de la Documentation* 4:

148-168, 1961;
and the debates on the UDC at the first session of the Unesco Advisory Committee on Bibliography, Documentation and Terminology (Paris, September 25-29, 1961), which was discussed by the representatives of 12 international and 10 national scientific organizations and bodies: *Bibliographie - Documentation - Terminologie* 5: 83-93, 1961.
On the work of the FID/C 3 Commission, see:
Arntz, H., 'Problematik einer Klassifikation der Sozialwissenschaften' [Problems of social science classification], *DK-Mitteilungen* 6: 21-23, Nov. 1962;
see also: Striganov, V., 'O rabote nad edinoi skheme klassifikatsiya dlya krupnykh universal'nykh bibliothek' [On the works on a uniform system of classification for large universal libraries], *Bibliotekar* 6: 29-32, 1962.
By its decree of May 11, 1962, the Council of Ministers of the Soviet Union made the use of the UDC system compulsory in libraries of science and technology: 'O merakh po uluchsheniyu organizatsii nauchnotekhnicheskoi informatsii v strane' [On improving the organization of scientific-technical information in the country].

41. Mölgaard-Hansen, R. and Rugby, M. (eds.), *Proceedings of the First Seminar on UDC in a mechanized retrieval system. Copenhagen, September 2-6, 1968.* Copenhagen, Danish Centre for Documentation, 1969; *FID/CR Report No. 9.* FID Publ. Serie No. 405;
Schmide, K. and Koch, K., *The use of the UDC in the production of mechanized indexes.* Frankfurt/M., Zentralstelle für machinelle Dokumentation, ZMD-A 21. 75 p.
42. Coblans, H., 'New methods and techniques for the communication of knowledge', *Unesco Bulletin for Libraries* 7: 154-175, 1957, a Unesco-commissioned study whose essential statements still hold to this date. This study has been thoroughly discussed, also in the Soviet Union: 'Soveshchaniya po novym metodam i tekhnike informatsii' [Discussions on the new methods and technique of information], *Sovetskaya Bibliografiya* 49: 58-61, 1958.
Volume I of the Washington Conference *(Op. cit.)* is largely devoted to these problems (p. 665-816).
For a comprehensive survey on this subject matter, see:
Vickery, *Techniques modernes de documentation; analyse des systèmes de recherche de documents.* Paris, 1962. 178 p.;
Samuelson, K. (ed.), *Mechanized information storage, retrieval and dissemination. Proceedings of the FID/IFIP Joint Conference.* Amsterdam, North-Holland Publ. Comp., 1968, 729 p.;
Henley, J., *Computer-based library and information systems.* London, Macdonald, 1970, 84 p.
43. Polushkin, V.A.: *Op. cit.*
44. 'UK Classification Research Group', *FID News Bulletin* 9, 1970, 106 p.;

see also *Klassifikationssysteme und Thesauri.* Frankfurt/M., 1969, 224 p.;
Barke, I.L., and Tschacke, L., *Thesauren und ähnliche Begrifflisten* [Description of 173 thesauri and lists of descriptors between 1960 and 1968]. Dresden, TU, 1969, 134 p.;
Aligned list of descriptors. Economic and social development, Vols. 1-5 + 1 guide. Paris, OECD, 1969. (Compiled by the 'documentary pool': ILO, FAO, OECD, Maison de Science de l'Homme, Deutsche Stiftung für Entwicklungsländern).

45. De Grolier, E., *Op. cit.* (n.36), and for the description of an operational mechanized information system in social sciences, see Thompson, G.K., Schieber, W.D. et al., *ISIS – A short Guide.* Geneva, ILO, LD/Notes/50. (An integrated set of information systems). See Appendix A of this volume.
46. Meyriat, J., *Rapport sur les techniques modernes de documentation dans les sciences sociales et leur application aux pays en voie de développement.* Paris, CIDSS Secr., 62/4, Novembre 1962.
47. *L'échange de résumés analytiques de documents scientifiques dans le domaine de l'économie appliquée.* Geneva, ONU, Commission Economique pour l'Europe, Febr. 20, 1963, 26 p. E/ECE/474;
Bibliographical Information on some of the existing Documentation Services. Joint ECE-Unesco Consultation, Working paper 2. Geneva, Jan. 15, 1963, 16 p.
48. For its activities, see: 'Le Comité International pour la Documentation des Sciences Sociales: Dix années d'activité' [ICSSD: A report on the first ten years]. *Extrait de la Revue internationale des Sciences Sociales,* Vol. XIV, 1 p. 18; see also periodic reports in the *Social Science Information.*
49. *International Bibliography of Sociology,* 1952;
International Bibliography of Political Science, 1954;
International Bibliography of Economics, 1955;
International Bibliography of Social and Cultural Antrhopology, 1957
50. Liebaers, H., Articles in *Unesco Bulletin for Libraries* 2,3: 1970.

NOTES TO 'SCIENTIFIC LIBRARY IN THE AGE OF THE SCIENTIFIC AND TECHNICAL REVOLUTION'

1. Data from the statistics of the Ministry of Education and from the *Bulletin of Science Organization* (Library of the Hungarian Academy of Sciences), Vol. 9 (1969), no 6.
2. In reality, the number of research works was less than that, however fragmented the projects were.
3. Marx, *Theories of surplus value,* Part I.

4. The number of professionals specializing in two or more disciplines had increased over the past few years, e.g. engineer-economists, applied and mathematical linguists, etc. The development of complex and borderline disciplines involved the complex training and qualification, too.
5. *Chemisches Zentralblatt* may be looked upon as one of the classical examples for the independence of documentation because it has been, ever since its start, an abstracting journal independent of any library. However, this proves nothing; nor does its reverse: had it been published within the framework of a library, this would not prove either that the natural workshop for the editing of an abstracting journal is the library.
6. Mikhaylov, A., Cerniy, A. and Gilyarevsky, R., *Osnovy informatika* [Fundamentals of informatics], 2. pererab. i dopoln. Izd. Moscow, Nauka, 1968; *An introductory course on informatics,* Unesco, COM/WS/147.
7. Akos, K., *A tudomány fejlödése és a könyvtárügy. A Egyetemi Könyvtár Evkönyvei* [The auvancement of science and librarianship], Vol. 2. Budapest, 1964, p. 103-107.
8. In all probability, it was the anxiety the over-growth of 'machine culture' that prompted László Németh, the outstanding writer, to refer in his work *Egetö Eszter* to the machine as *a devastating marauder.*
9. Line, M., *Automation and national libraries;*
   Foskett, D.J., *The changing role of the librarian;*
   Brookes, B.C., 'Knowledge versus information', *Times Literary Supplement,* Jan. 15, 1971.
10. This would require a series of analytical studies which would tackle the problems connected with the possibilities of publication. The 'publicity' of the theme should involve an open discussion on the major theoretical problems of scientific librarians, along with some more important information problems of the individual branches of science in organs other than the professional library journals. (There have been some initiatives taken in this respect, see e.g.n.9.
11. Lectures on this subject of Snow *(The two cultures and a second look: An expanded version of the two cultures and the scientific revolution. New York, 1964).* and Russel provoked debates all over the world dealing with the free time ('time-budget'), the division of labour and the perspectives of the personality's development, together with debates on 'alienation' connected, directly or indirectly, with the 'two cultures'.
12. Marx's drafts on the problems of political economy, first published in *Bolshevik* 11/12, 1938.
13. Commissioned by Unesco's European Coordination Center for Research and Documentation in Social Sciences (Vienna) and by the International Association of Sociology, Prof. Alexandre Szalai was one of the leaders of a major East-West research project on the problems of 'time-budget'. *The use of time. A cross-national comparative survey of*

daily activities of urban and sub-urban populations in twelve countries. The Hague – Paris, Mouton (Publications of the Vienna Centre, Vol. 5), 1972, ca. 880p.

14. Reference to a passage in Imre Madách's classical work *The Tragedy of Man,* Scene 12: '. . . To keep you wide awake, you'll kneel on peas.'; Plato: 'I'll dream of beauties even on my knees . . .'

NOTES TO 'AN OUTLINE OF THE SYSTEMATIZATION OF SCIENCE ORGANIZATION'

1. *World Marxist Review* 1961,7 ('The crisis and the workers'; the part entitled 'The corrections of the scientific-technical revolution' from the contribution by Barel, Y. and Menshikov, S., 'What about the World Cycle? ').
2. For a more recent Marxist appraisal of the impact of the scientific and technical revolution on society, see: *Sciences Sociales d'Aujourd'hui* 1: Moscow, 1969, a collection of papers entitled in the special number *La révolution scientifico-technique* of the Academy of Sciences SSSR, especially the papers of A. Roumiantsev, A. Akhiezer, V. Maratov, and Y. Melechtchenko.
3. *World Marxist Review* 1961,10 (the part on 'Science and the future of mankind' – including 'Scientific and social justice') and *World Marxist Review* 1961,12 by Bernal, J.D., the part on 'The technical revolution and social progress', in particular. Other material on this problem in the same *Review* (4, 1963): 'The function of science in modern society' (Review).
4. Sub-chapter 1, Chapter I, Part II of the Programme. 'Programme du PCUS'. *Cahiers du Communisme* 1961, Dec., numéro spécial. In *Grundrisse zur Kritik der politischen Okonomie* (Berlin, 1953. p. 586-587) Marx foresaw this. (French edition: *Fondements de la critique d'économie politique.* Paris, Ed. Anthropos, 1968).
5. Richta, Radovan et al., *Civilization at the crossroads. Social and human implications of the scientific and technical revolution.* 3rd ed. enl. New York, International Arts and Sciences Press, 1969, 37 p. (French ed.: *La civilisation au carrefour.* Paris, Ed. Anthropos, 1969, 463 p.); Salomon, J.-J., *Science et politique.* Paris, Ed. du Seuil, 1970, 406 p.
6. There are many monographs, articles and studies on national science policy, for instance large monographs such as Gilpin, R., *La science et l'Etat en France.* Paris, Gallimard, 1970, 412 p.; or *The social, economic and organizational problems of the Soviet science.* Table of contents, introduction, summaries of the chapters. Moscow, The Concrete Social Research Institute of the USSR Academy of Sciences, 1970, 107 p.

7. Spaey, J. et al., *Le développement par la science. Essai sur l'apparition et l'organisation de la politique scientifique.* Paris, Unesco, 1969, 204 p.
8. Ossowska, M. and Ossowski S., 'The science of science', *Organon (Warsaw)* 1: 1-12, 1936, and republished in the same review: 1: 1963; see also for a contemporary and comprehensive approach, the collected papers in the volume *Science of science.* Published under the auspices of the International Body of the Science of Science Foundation (London). London/Toronto, M. Goldsmith, 1966. For a Marxist approach of this problem, Dobrov, G.M., *Nauke o nauke* [The science of science: Introduction to the general study of science]. Kiev, Naukova Dumka, 1966.
9. Cf. Salomon, J.-J., *Op. cit.* (n 5), p. 375.
10. Working paper presented by the 23rd session on the revision of UDC class 3 (social sciences), Groningen, October 1970, FID C/3 70-30.
11. Bernal, J.D., *Science in history.* London, 1954.
12. *Etude bibliographique comparative sur les dépenses consacrées à la recherche scientifique et technique de 1957 à 1963.* Paris, April 30, 1963, 25 p. Unesco/ROU/24.
13. *World without war.* London, 1959.
14. 'National science policies in Europe'. *Science policy studies and documents* 17: 45,46,48, 1970 (Unesco); see also from the same serial 18: 1970, 'The role of science and technology in economic development'.
15. *Le développement par la science* (cf. n7), p. 70-71.
16. Hill, D.W., *Research as a business. Transactions of the Manchester Statistical Society. Session 1957-1958.* Lockwood, Norbury, 1958. p.
17. Russo, F. and Erbes, R., 'La recherche-développement. Concepts et problèmes de base, environnement et propagation de l'information scientifique et technique'. *Cahiers de l'ISEA* T/1, Paris, 1959, 80 p.
18. 'The role of science in the modern society', *World Marxist Review* 4: 1963.
19. Marx, *Theories of surplus value,* Part I.
20. Gore, G., *The scientific basis of national progress, including that of morality.* London, Cass, 1970. With index. First published 1882.
    Hole, I., *An essay on the history and management of literary, scientific mechanics' institutions and especially how far they may be developed.* London, Cass, 1970. With index. First published 1853.
    Proctor, R.A., *Wages and wants of science-workers.* London, Cass, 1970. With index. First published 1876.
21. *Current trends in scientific research.* New York/Paris, UN-Unesco, 1961, 246 p.
22. *Main trends of research in the social and human sciences. I: Social sciences.* The Hague/Paris, Mouton/Unesco, 1970, xlviii + 819 p.
    Tendances principales de la recherche dans les sciences sociales et humaines. I: Sciences sociales. The Hague-Paris, Mouton/Unesco, 1970, lii + 987 p.

23. UN-ECOSOC Advisory Committee on the Application of Science and Technology to Development: *World plan of action for the application...* New York/Paris, 1970. E/AC. 52/R.13. (SC/WS/342, ROU/203 Unesco).

NOTES TO 'THE POINTS OF DIVERGENCE BETWEEN RESEARCH AND SCIENTIFIC INFORMATION'

1. Bóna, E. and Farkas, J., 'A tudomány jelenlegi struktúrájanak és fejlödésének néhány ellentmondása' [Some contradictions of the present structure of science and of its developments], *Magyar Tudomány* 6: 435-446, 1970.
2. For the most recent synthesis of this thematic group see: Mikhaylov, A., Cerniy, A. and Gilyarevski, R., *Osnovy informatika* [Fundamentals of informatics], 2. pererab. i dopoln, izd. Moscow, Nauka, 1968.
3. Marx, Gy., *Társadalmi Szemle* 3 : 53-61, 1970.
4. Marx, K., *Theories on the surplus value,* Part 1. London, Lawrence and Wishart, 1970.
5. Marx, *Op. cit.* p. 373.
6. Besterman, T., *A world bibliography of bibliographies.* 3rd. ed., Vols. 1-4. Geneva, Societas Bibliographica, 1955/1956.
7. *Inventaire raisonné des services périodiques de documentation des sciences sociales.* Paris, 1951. 140 p.
8. Koblitz, J., 'Die erhöhte Bedeutung der Dokumentation und Information' [The enhanced significance of documentation and information], *Dokumentation* 2: 33-41, 1962.
9. Kofnovec, L., 'Das Veralten wissenschaftlich-technischer Informations' [The redundancy of scientific-technical information], *Dokumentation* 2: 42-49, 1962).
10. Kosel, G., *Die socialistische Enzyklopädie als Mittel einer grundlegender Verbesserung der Gemeinschaftarbeid in Forschung und Entwicklung* [Socialist encyclopaedia as a fundamental means of improving collective research and development work]. Berlin, Deutsche Bau-Enzyklopädie, 1960, p. 22.
11. Engels, *The situation of England. An outline of a criticism of the national economy.*
12. Engels, *Works.*
13. *Scientific manpower for applied research.* Paris, OECD, 1956. 27 p.
14. The European translation 'pool' has been organized and is being operated at the Technical University of Delft with US and Canadian participation.

15. De Grolier, E., 'La Conférence Internationale sur l'Information scientifique, Washington', *Bulletin des Bibliothèques de France* 1: 3-10, 1959.
16. In addition to the existing ones, reference services are being organized in growing numbers in the field of social sciences; *The Journal of Economic Abstracts* was started in 1963 after the large-scale investigation, undertaken by the American Economic Society, had come to the conclusion that such a periodical would be indispensable. This is the more noteworthy as the periodical of the Society, the *American Economic Review*, had already been devoting half of its pages to reviews, bibliographies and abstracts. Four bulky volumes with an index of selected economic periodicals published between 1886 and 1954 had by then also been published again under the aegis of the Society (Homewood, Ill., 1962.)
    In a small country as for example Hungary, the National Documentation Committee found that some fifty publications on economic documentation are being edited by more than twenty institutions.
    The French *Fichier Bibliographique de l'Entreprise* (a quarterly for abstracts in enterprise economy) and the journal *Interproductivité* stressing enterprise documentation, or such periodicals as the *Byulletin Zarubezhnoi Ekonomicheskoi Informatsii* [Foreign Economic Information Bulletin], published by the Research Institute of the State Planning Office of the Soviet Union, were also started in the early Sixties.

The author wishes to express his gratitude to Mr. Pàl Kelemen (London) for his help in editing the English manuscript of this book.

Bei Fragen zur Produktsicherheit wenden Sie sich bitte an:
If you have any questions regarding product safety,
please contact:

Walter de Gruyter GmbH
Genthiner Straße 13
10785 Berlin
productsafety@degruyterbrill.com